THE NATIONAL ACADEMIES
Advisers to the Nation on Science, Engineering, and Medicine

The **National Academy of Sciences** is a private, nonprofit, self-perpetuating society of distinguished scholars engaged in scientific and engineering research, dedicated to the furtherance of science and technology and to their use for the general welfare. Upon the authority of the charter granted to it by the Congress in 1863, the Academy has a mandate that requires it to advise the federal government on scientific and technical matters. Dr. Bruce M. Alberts is president of the National Academy of Sciences.

The **National Academy of Engineering** was established in 1964, under the charter of the National Academy of Sciences, as a parallel organization of outstanding engineers. It is autonomous in its administration and in the selection of its members, sharing with the National Academy of Sciences the responsibility for advising the federal government. The National Academy of Engineering also sponsors engineering programs aimed at meeting national needs, encourages education and research, and recognizes the superior achievements of engineers. Dr. Wm. A. Wulf is president of the National Academy of Engineering.

The **Institute of Medicine** was established in 1970 by the National Academy of Sciences to secure the services of eminent members of appropriate professions in the examination of policy matters pertaining to the health of the public. The Institute acts under the responsibility given to the National Academy of Sciences by its congressional charter to be an adviser to the federal government and, upon its own initiative, to identify issues of medical care, research, and education. Dr. Harvey V. Fineberg is president of the Institute of Medicine.

The **National Research Council** was organized by the National Academy of Sciences in 1916 to associate the broad community of science and technology with the Academy's purposes of furthering knowledge and advising the federal government. Functioning in accordance with general policies determined by the Academy, the Council has become the principal operating agency of both the National Academy of Sciences and the National Academy of Engineering in providing services to the government, the public, and the scientific and engineering communities. The Council is administered jointly by both Academies and the Institute of Medicine. Dr. Bruce M. Alberts and Dr. Wm. A. Wulf are chair and vice chair, respectively, of the National Research Council.

www.national-academies.org

ORGANIZING COMMITTEE FOR THE WORKSHOP ON NATIONAL SECURITY AND HOMELAND DEFENSE

JOHN L. ANDERSON, Carnegie Mellon University, Co-Chair
JOHN I. BRAUMAN, Stanford University, Co-Chair
JACQUELINE K. BARTON, California Institute of Technology, Steering Committee Liaison
MARVIN H. CARUTHERS, University of Colorado
LIANG-SHIH FAN, Ohio State University
LARRY E. OVERMAN, University of California, Irvine
MICHAEL J. SAILOR, University of California, San Diego
JEFFREY J. SIIROLA, Eastman Chemical Company, Steering Committee and BCST Liaison

Staff

JENNIFER J. JACKIW, Program Officer
CHRISTOPHER K. MURPHY, Program Officer
SYBIL A. PAIGE, Administrative Associate
DOUGLAS J. RABER, Senior Scholar
DAVID C. RASMUSSEN, Program Assistant

CHALLENGES FOR THE CHEMICAL SCIENCES
IN THE 21ST CENTURY

NATIONAL SECURITY & HOMELAND DEFENSE

COMMITTEE ON CHALLENGES FOR THE CHEMICAL SCIENCES
IN THE 21ST CENTURY

BOARD ON CHEMICAL SCIENCES AND TECHNOLOGY

DIVISION ON EARTH AND LIFE STUDIES

NATIONAL RESEARCH COUNCIL
OF THE NATIONAL ACADEMIES

THE NATIONAL ACADEMIES PRESS
Washington, D.C.
www.nap.edu

THE NATIONAL ACADEMIES PRESS • 500 Fifth Street, N.W. • Washington, D.C. 20001

NOTICE: The project that is the subject of this report was approved by the Governing Board of the National Research Council, whose members are drawn from the councils of the National Academy of Sciences, the National Academy of Engineering, and the Institute of Medicine. The members of the committee responsible for the report were chosen for their special competences and with regard for appropriate balance.

Support for this study was provided by the American Chemical Society; the American Institute of Chemical Engineers; the Defense Advanced Research Projects Agency under Contract No. MDA972-01-M-0001; the Camille and Henry Dreyfus Foundation, Inc. under Contract No. SG00-093; the National Institute of Standards and Technology under Contract No. NA1341-01-W-1070; the National Institutes of Health/National Cancer Institute under Contract No. N0-OD-4-2139; the National Science Foundation under Contract No. CTS-9908440; the U.S. Department of Energy/ Basic Energy Science under Contract No. DE-FG02-00ER15040; the U.S. Department of Energy/ Office of Industrial Technologies under Contract No. DE-AT01-01EE41424; and the U.S. Environmental Protection Agency under Contract No. CR828233-01-0.

All opinions, findings, conclusions, or recommendations expressed herein are those of the authors and do not necessarily reflect the views of the organizations or agencies that provided support for this project.

International Standard Book Number 0-309-08504-7

Additional copies of this report are available from:

The National Academies Press
500 5th Street, N.W.
Box 285
Washington, DC 20055
800-624-6242
202-334-3313 (in the Washington Metropolitan Area)
http://www.nap.edu

Printed in the United States of America

COMMITTEE ON CHALLENGES FOR THE CHEMICAL SCIENCES IN THE 21ST CENTURY

RONALD BRESLOW, Columbia University, Co-Chair
MATTHEW V. TIRRELL, University of California, Santa Barbara, Co-Chair
JACQUELINE K. BARTON, California Institute of Technology
MARK A. BARTEAU, University of Delaware
CAROLYN R. BERTOZZI, University of California, Berkeley
ROBERT A. BROWN, Massachusetts Institute of Technology
ALICE P. GAST,[1] Stanford University
IGNACIO E. GROSSMANN, Carnegie Mellon University
JAMES M. MEYER,[2] E. I. du Pont de Nemours and Company
ROYCE W. MURRAY, University of North Carolina, Chapel Hill
PAUL J. REIDER, Amgen, Inc.
WILLIAM R. ROUSH, University of Michigan
MICHAEL L. SHULER, Cornell University
JEFFREY J. SIIROLA, Eastman Chemical Company
GEORGE M. WHITESIDES, Harvard University
PETER G. WOLYNES, University of California, San Diego
RICHARD N. ZARE, Stanford University

Staff

JENNIFER J. JACKIW, Program Officer
CHRISTOPHER K. MURPHY, Program Officer
SYBIL A. PAIGE, Administrative Associate
DOUGLAS J. RABER, Senior Scholar
DAVID C. RASMUSSEN, Project Assistant

---

[1] Committee member until July 2001; subsequently the Board on Chemical Sciences and Technology (BCST) liaison to the committee in her role as co-chair of the BCST.

[2] Meyer's committee membership ended March 2002, following his retirement from DuPont.

Preface

Initially, the Challenges for the Chemical Sciences in the 21st Century project was designed to be a series of five workshops encompassing the main technological areas to which the chemical sciences contribute. After the events of September 11, 2001, it was recognized that chemists and chemical engineers have always contributed significantly to our nation's defense capabilities and that now they will play an increasingly important part in homeland security. Thus, the National Security and Homeland Defense Workshop was arranged on an emergency basis. It is hoped that the presentations and discussions at the workshop found in this report will help chemical scientists understand how their research can be applied to national security problems and will guide them in new directions to ultimately enhance the safety of U.S. civilians and military personnel.

In some cases, modified or improved existing technologies were identified as likely contributors to national security solutions, and in others, completely new technologies were called for. However, the workshop was not designed to provide specific recommended solutions to national security and homeland defense problems (see Appendix A for the Statement of Task). The workshop report is just that—a report of the proceedings of and discussions at the workshop that focused on research in chemistry and chemical engineering clearly related to national security. Topics in other areas of the chemical sciences are addressed in the other reports in the "Challenges" series including *Beyond the Molecular Frontier: Challenges for Chemists and Chemical Engineers* as well as in *Making the Nation Safer: The Role of Science and Technology in Countering Terrorism.*

Following the workshop, the organizing committee met to reach preliminary consensus on the workshop findings and to create an outline for its report. The report was fully developed through iterations among the committee members. In addition to summaries of the speaker presentations (Appendix D), the committee has attempted to capture participants' input from both the plenary and breakout sessions in the report chapters. Illustrative comments from presentations and subsequent discussions are highlighted in boxes interspersed throughout the chapters.

This study was conducted under the auspices of the NRC's Board on Chemical Sciences and Technology with assistance provided by its staff. The committee acknowledges this support.

John L. Anderson and John I. Brauman
Co-Chairs
Organizing Committee for the Workshop on
National Security and Homeland Defense

Acknowledgment of Reviewers

This report has been reviewed in draft form by individuals chosen for their diverse perspectives and technical expertise, in accordance with procedures approved by the NRC's Report Review Committee. The purpose of this independent review is to provide candid and critical comments that will assist the institution in making the published report as sound as possible and to ensure that the report meets institutional standards for objectivity, evidence, and responsiveness to the study charge. The review comments and draft manuscript remain confidential to protect the integrity of the deliberative process. We wish to thank the following individuals for their review of this report:

Peter K. Dorhout, Colorado State University
Catherine C. Fenselau, University of Maryland
Nancy B. Jackson, Sandia National Laboratories
C. Bradley Moore, Ohio State University
George Parshall, E. I. duPont de Nemours and Company (retired)
K. John Pournoor, 3M
Steven J. Sibener, University of Chicago
George M. Whitesides, Harvard University
Charles Zukoski, University of Illinois

Although the reviewers listed above have provided many constructive comments and suggestions, they were not asked to endorse the conclusions or recommendations nor did they see the final draft of the report before its release. The review of this report was overseen by R. Stephen Berry, University of Chicago. Appointed by the National Research Council, he was responsible for making certain that an independent examination of this report was carried out in accordance with institutional procedures and that all review comments were carefully considered. Responsibility for the final content of this report rests entirely with the authoring committee and the institution.

Contents

Executive Summary

The public has long recognized the potential benefits of research in chemistry and biology to the health and welfare of our society, and the relevance of that research to national security has become clearer since September 11, 2001. Chemical and biological agents can be a realistic threat to the safety of our civilian and military personnel. The nation must address these threats in the context of threat reduction, preparation, response, and neutralization and remediation. Technology alone cannot meet the goals of security and safety that our populace desires: the United States leads the world with its experts and education in the chemical sciences and these should be deployed to help us reach our security goals. We must not only incorporate current knowledge in chemistry and chemical engineering, but also engage our nation's experts in advancing the frontiers of knowledge through research to minimize the chance of a terrorist attack, minimize the extent of injury and material damage should such an attack take place, and remedy the effects of such an attack.

As a response to the events of September 11, 2001, the National Security and Homeland Defense Workshop was included as one of six workshops held as part of "Challenges for the Chemical Sciences in the 21st Century." The workshop topics reflect areas of societal need—materials and manufacturing, energy and transportation, national security and homeland defense, public health, information and communications, and environment. The charge for each workshop was to address the four themes of discovery, interfaces, challenges, and infrastructure as they relate to the workshop topic:

- Discovery—major discoveries or advances in the chemical sciences during the past several decades.

- Interfaces—interfaces that exist between chemistry/chemical engineering and such areas as biology, environmental science, materials science, medicine, and physics.
- Challenges—the grand challenges that exist in the chemical sciences today.
- Infrastructure—infrastructure that will be required to allow the potential of future advances in the chemical sciences to be realized.

The National Security and Homeland Defense workshop serves two main goals. First, it provides a mechanism to broadly engage the chemical sciences R&D community in assessing the needs and recognizing the opportunities for R&D in this important area. Second, it provides federal agencies with information that can guide and support their investment in R&D in key areas where the chemical sciences can make a unique contribution. This workshop report is complementary to *Beyond the Molecular Frontier: Challenges for Chemistry and Chemical Engineering*, the report on the future of the chemical sciences that is being produced by the Steering Committee for the "Challenges" study chaired by Ronald Breslow and Matthew V. Tirrell, and *Making the Nation Safer: The Role of Science and Technology in Countering Terrorism*, the report produced by the National Research Council committee chaired by Lewis M. Branscomb and Richard D. Klausner that makes specific recommendations on courses of action to be taken to prepare our nation for potential terrorist attacks.

On January 14-16, 2002, the Workshop on National Security and Homeland Defense was held to identify both the challenges faced as a result of terrorist threats and the most effective means available for chemists and chemical engineers to respond to those challenges. The chemistry and chemical engineering community at large was invited to attend; those who came did so of their own volition and their interest in the subject matter and service to society. While these participants were individuals with expertise in the chemical sciences from industry, academia, and the federal government, they were not necessarily experts on national security and homeland defense issues. Furthermore, by holding an unclassified workshop, information on some existing or developing technologies could not be included in the discussions. Nevertheless, approximately 100 workshop participants were involved in stimulating presentations and discussions on the use of chemical sciences to meet the needs of national security and homeland defense. The ideas and challenges identified in these presentations, as well as the four themes listed above, were used as a starting point for breakout sessions, where the participants offered their thoughts and expanded on technical concepts and needs identified in the lectures.

Material from both the presentations and the breakout sessions was used by the committee as the basis for the findings drawn in this report. After an introductory breakout session that identified ways that the chemical sciences have contributed to national security and homeland defense throughout the past few decades, the goal of the committee was to identify a series of grand challenges, the overarching needs and goals for effectively using the chemical sciences in the counterterrorism effort. Each of the grand challenges is composed of a series of

Some Grand Challenges for Chemistry and Chemical Engineering

- Energy independence—to reduce our reliance on foreign nations. Opportunities in the chemical sciences include the development of more efficient engines, better materials for energy storage, and alternative energy sources. More detailed information can be found in the workshop report on Energy and Transportation, part of the "Challenges" series.
- Supply chain safety—to protect the products and services on which our economy and society depend. The opportunities for chemical scientists include improved security plans at chemical production sites, automated detection systems that can sense agents inside of closed containers, and improved placarding of railroad tank cars.
- Secure chemistry—to minimize the misuse of chemicals and chemical assets as weapons of destruction. Challenges for chemical scientists include the substitution of less toxic chemicals in the manufacturing process and the minimization of the accumulation and storage of hazardous intermediates and products.
- Integrated response measures—to minimize loss of life and property. The challenges for chemical scientists include the development of detectors and information systems that can provide responses for chemical, biological, and radiological weapons to maximize the effectiveness of a diverse group of responders.
- Chemical, biological, and radiological detectors—to minimize or prevent harm. Opportunities in the chemical sciences range from the development of rugged, portable detectors for field use to better methods of sample preparation.
- Collective and personal protection systems—to avert infection or contamination. Challenges for chemical scientists include improved air filtration systems and the effect of weather patterns on agent dispersal.
- Immediate and extended medical countermeasures—to diagnose and treat infection and contamination. The opportunities for the chemical sciences include gaining a better understanding of virus replication and better computer modeling for drug design.
- Neutralization, decontamination, and disposal procedures—to safely return infrastructure and property to use. Challenges for chemical scientists range from the development of environmentally benign decontaminants to the surface science of agents and decontaminants.

more specific challenges or barriers that must be overcome to meet the goal stated in the grand challenges. Likewise, each specific challenge has related research needs, which, once met, can aid in overcoming the scientific barriers to the grand challenges.

The committee's findings to guide research on defense-related topics are listed below.

1. Extensive research opportunities exist throughout the many facets of chemistry and chemical engineering that can assist national security and homeland defense efforts.

  a. Research in systems and analysis is required to apply new knowledge of basic science to developing tools and products for national security and homeland defense. Three major needs were identified at the workshop:

  - **Detectors with the broadest possible range for chemical and biological agents.** New sensing and sampling technologies will yield improved threat detection and identification technologies. Specific detectors that provide speed, high sensitivity, and great reliability are essential.

  - **Miniaturization of detection and identification equipment for use in the field.** This is needed by soldiers, first responders, and other emergency and military personnel. This research will also allow the development of detection systems on a chip. Understanding materials and transport phenomena at the micron scale may facilitate research on miniaturized processes.

  - **Personal and collective protection measures.** Research is required to determine how to use new materials and methods to develop protective clothing, new drugs and vaccines, and infrastructure protection (air filters, stronger construction materials, decontaminants).

  b. Addressing research opportunities in manufacturing would better prepare the nation for a terrorist attack.

  - **Minimization of the hazards of chemical transport and storage.** Chemical transport routes by rail, truck, or barge pass through population centers and neighborhoods. Large storage tanks of chemicals often exist at industrial sites. The inherent danger of chemical transportation and storage can be reduced if new manufacturing processes are developed that use less reactive, less harsh chemicals or produce and consume the dangerous chemicals only on demand and on site.

  - **Rapid production scale-up procedures.** If these procedures are established before they are needed, they can be implemented immediately in an attack. Although investigation into scale-up of all stages of production should occur, the scale-up of separation and purification is a specific challenge.

c. Fundamental research in basic science and engineering is essential to meet these needs. Much basic research stems from the need to solve specific problems faced by society. Supporting fundamental research for national security and homeland defense will enable discovery and understanding in many areas of chemistry and chemical engineering. These areas include materials development, pharmaceutical modes of action, particulate air transport and dispersal, catalysis, chemical and biochemical binding, solids flow and crystallization, energy sources and batteries, and sampling methodology and preparation. Research in many of these areas will be needed to meet the challenges outlined above.

2. Infrastructure enhancements are needed to promote the success of research in national security and homeland defense.

- **a. Access to secure facilities.** Many important research efforts will require the use of hazardous materials, classified data, and specialized facilities. Collaboration with chemists, chemical engineers, and other scientists at government facilities will be necessary.
- **b. Transformation of graduate education.** Changes in the infrastructure of graduate education to remove barriers to interdisciplinary and multi-disciplinary research will enable students to expand their range of knowledge and to learn how to work in concert with others to achieve maximum effectiveness. This will be needed for them to participate successfully in any research program after graduation, especially one centered on national security and homeland defense related issues.
- **c. Instrumentation for research.** Much of the needed research will depend strongly on new instrumentation to replace existing obsolete equipment as well as to allow new experiments. This includes major, multi-user equipment as well as specialized instrumentation for individual investigators.

1

Introduction

The September 11, 2001 terrorist attacks on New York City and Washington, D.C. clearly indicated that massive, coordinated acts intended to cause enormous destruction and loss of life could be carried out within the borders of the United States. The response by the U.S. government to combat future attacks of this nature has been extensive, not only from the military but also from a vast number of agencies within the federal government.

Unfortunately, the anthrax attacks throughout the eastern United States subsequent to September 11 demonstrate an additional component of the terrorist threat facing our nation. Prevention of a conventional attack using high explosives (or in the case of September 11, jet fuel) is not enough. The ability to detect chemical, biological, or radiological toxins before an attack and rapidly and effectively decontaminate facilities and treat individuals should such attacks occur is also a key component in combating terrorism. The possibility of these attacks is no longer considered remote, and our chemists and chemical engineers are presently engaged in research that will strengthen our nation's security and homeland defense capabilities.

Chemical and biological agents present a "dual-use" dilemma.[1] A facility needed to develop vaccines against a biological agent could also be used to produce the agents for warfare. However, the dual-use principle can be turned to advantage; for example, the science and engineering needed for discovery and

---

[1]D. R. Franz. 2002, Current Thought on Bioterrorism: The Threat, Preparedness, and Response. Presentation, Workshop on National Security and Homeland Defense, Irvine, CA. (See Appendix D.)

Scientific Success Requires Public Support
K. John Pournoor
3M Company

In the past century, much scientific work has been accomplished because the public put its support behind particular initiatives. For instance, from the 1940s through the 1960s, the public wanted to overcome the morbidity and mortality associated with infectious diseases, so our entire scientific system geared toward funding the development of the antibiotics that we have today. In the 1970s, the American public supported the fight against cancer, in the 1980s the battleground was viral diseases, especially HIV, and in the 1990s, the focus was on understanding the human genome. Each of these scientific accomplishments was achieved because there was at least a decade or more of sustained focused activity and publicly supported funding.

In this decade, science is and will be focused on developing a repertoire of prevention, detection, and intervention capabilities, as long as the public support for these activities remains strong. To put the fruits of this labor into the hands of the public will require both government and industry support as well as high efficacy, reasonable cost, reasonable complexity, and availability on a routine, everyday basis. So the list of priorities for scientific research related to national security and homeland defense needs to be amplified and channeled in many different directions by academia, government, and industry, to create an overarching cohesive approach.

scale-up of pharmaceutical agents in response to an attack could also be applied to improve the health of our residents in normal times, and vice versa.

Indeed, basic discoveries and inventions in the chemical sciences over the past two decades have been applied to the national security and homeland defense arena. One highly developed branch of science is analytics. Much of the instrumentation developed is being used to detect harmful chemicals and biological agents. For example, mass spectrometers have been fielded by the Department of Defense (DOD) for more than 15 years to monitor chemical and biological agents. Also, in the past 6 years, photon-based detection techniques have been combined with mass spectrometry, and DOD and the Defense Advanced Research Project Agency are heavily supporting development of mobile units for detection of both chemical and biological agents in the field. The development of aerosol time-of-flight mass spectrometry for on-line continuous monitoring of aerosol particles, including the chemistry between particle and gas, was originally developed[2] for environmental science purposes but could be applied to detect military or terrorist attacks through an airborne delivery system. Other analytical developments with current or potential application to national security are advanced spectroscopic techniques for characterization of volatile and semi-volatile particulates in the atmosphere, microfluidics, chemical signature detection, single particle monitoring, sensor arrays, and high throughput screening. A challenge is to extend this example to other basic discoveries in chemistry and to invent new methodologies for mobile detection.

Materials development and synthesis is another important dual-use type of chemistry. Developments over the past few decades include a number of electronic materials and their processing, fuel cells and batteries, photoresist and semiconductor synthesis, high-performance composites (structural components) and nanocomposite materials, colloidal nanoparticle technology, solid-state lasers, and light-emitting diodes.

Advances in biotechnology have been tremendous. There are many issues in biochemistry and biotechnology that have not yet been completely elucidated; however, progress has been made. For example, antibiotics have evolved, the polymerase chain reaction has been developed, and DNA sequencing, genomics, proteomics, combinatorial synthesis, and selective complexation and recognition chemistry have all advanced.

All of these discoveries of chemistry, chemical engineering, and other collaborative sciences would not have been possible without new synthetic methods and improved theory and modeling capabilities.

Research opportunities for the chemical sciences in combating chemical warfare, biological warfare, and general terrorism are vast. These opportunities can be classified into the following basic processes:

- Basic science
- Systems and analysis
- Manufacturing

The skills and knowledge of those in all chemical and chemical engineering disciplines are needed for these processes. Additionally, collaboration among the disciplines must be cultivated—especially between researchers who synthesize new chemicals and materials and those who develop new processes to manufacture them. There must be parallel integration of fields, so that discoveries quickly lead to useful technology. In a terrorist attack, a quick response is essential, whether it be decontamination of a site or scale-up of drug or vaccine production.

---

[2]K. A. Prather. 2002, Overview of Real-Time Single Particle Mass Spectrometry Methods. Presentation, Workshop on National Security and Homeland Defense, Irvine, CA. (See Appendix D.)

BASIC SCIENCE

Chemistry and chemical engineering have much to contribute to many aspects of national security and homeland defense. Some of these contributions are obvious and in many cases already underway. These are areas in which the basic science is largely in hand and development is the next important step.

There remain, however, fundamental issues that depend on advances in basic science. As the workshop has revealed, there are many aspects of our basic knowledge that must be improved in order for us to solve these problems. For example, our ability to detect specific chemicals, or mixtures of chemicals, without false positive responses will require improvements in our understanding and implementation of many basic spectroscopies in a variety of frequency regions. Knowledge of the spectroscopic signatures of common environmental interferences will also require improvement. Miniaturization of equipment will depend on new materials and new ways of handling small amounts of materials. This is true both in the chemical and biological sectors. Our response to attacks depends on our knowledge of how chemicals interact with decontaminating agents and the development of new methodologies for cleanup that solve the decontamination problem without significant human and environmental degradation. Finally, our ability to anticipate new threats and problems depends critically on our understanding of what is possible and what we can do to prevent attacks. For example, new methods of manufacturing that make use of nontoxic materials are critical. Similarly, the synthesis of new drugs, antidotes, and other pharmaceuticals is clearly necessary, and if attacked, faster, more efficient clinical testing and regulatory approval might be needed.

The general societal benefits of all of these basic advances are substantial. Indeed, it can be argued that the overall value to our society will extend well beyond national security issues and will far exceed the initial investments, because many of these advances will result in improved quality of life and health for all of society.

SYSTEMS AND ANALYSIS

Miniaturization—chemistry and biology on a chip—has resulted only because of the confluence of science and engineering. The development of microfluidics technology has mainly been driven by the need to miniaturize, integrate, and automate biochemical analysis to increase speed and reduce costs.[3] We are experiencing a revolution in the miniaturization of chemical systems for detection and analysis of hosts of chemical and biological materials and agents. Applications of basic principles of electrokinetics, hydraulics, and surface science have

---

[3]A. W. Chow. 2002, Microfluidics: Development, Applications, and Future Challenges. Presentation, Workshop on National Security and Homeland Defense, Irvine, CA. (See Appendix D.)

permitted advances in microfluidics for moving nanovolume fluid elements to targeted destinations for chemical reaction and analytical chemistry. The complex chemical and physical processing that can now be done on a square-centimeter chip required an entire laboratory 10 years ago.

The use of detectors in the field, whether military or civilian, demands more robust, smaller, lighter, and lower power instrumentation than laboratory use requires. Field instruments will require high levels of redundancy and orthogonality in order to provide the users with high confidence. The term "orthogonality" here refers to the existence of different types of sensors in the same instrument targeted to the same analyte, whose operating principles are so different that they are unlikely to provide a false positive signal. False positive readings lead to a loss of confidence in the instrument, sometimes with disastrous consequences.[4] Spectroscopic sensing devices will need to depend on multivariate sensing, comparing specific, targeted portions of the spectrum of the detected molecule to similar portions of the target molecule spectrum to overcome multiple environmental interferences. Although not important for laboratory instruments, field instruments will require real-time, or instantaneous, response. Reliable on-line detectors will require multiple schemes of analysis and associated data handling that are uncorrelated through both model and measurement. This is a new paradigm for detection.

The need for improved or new technologies for a multitude of purposes exists. For practical use, these technologies must be incorporated into a larger instrument containing other components. Systems integration will be of great importance as science progresses with national security and homeland defense. In addition to the discussion in this report, information on this topic can be found in Chapter 10: Complex and Interdependent Systems of *Making the Nation Safer: The Role of Science and Technology in Countering Terrorism*.[5]

MANUFACTURING

The creation and scale-up of processes to make sufficiently large quantities of drugs, vaccines, and other biological agents for testing and distribution challenges the creativity and knowledge of scientists and engineers. Here it is crucial that synthetic, physical, and analytical chemists, and process chemical engineers work in parallel to not only achieve the desired yield and purity, but also to greatly reduce the time from discovery to production. The emphasis of pharmaceutical production will be on continuous processing. Today, the timeline from discovery to scale-up of a pharmaceutical agent is 1 to 3 years (not including clinical trials);

---

[4]D. H. Stedman. 2002, A Skeptical Analysis of CBW Detection Schemes. Presentation, Workshop on National Security and Homeland Defense, Irvine, CA. (See Appendix D.)

[5]National Research Council. 2002. Making the Nation Safer: The Role of Science and Technology in Countering Terrorism. Washington, D.C.: National Academy Press.

Continuous Quality Control
Mauricio Futran
Bristol-Myers Squibb

Currently, testing of an operation (fermentation, reaction, separation, etc.) in the pharmaceutical industry requires taking a sample, bringing it to the laboratory, running the test, and analyzing the results. Only then is the operation approved or rejected. Not only is that process inefficient for the industry in general, but it also prohibits the rapid development, scale-up, and production of drugs in a national emergency. Instead, the industry needs continuous quality verification: it needs the controls and measurements of the process integrated into the process so that the quality of the product is known in real-time. Instituting this change will call for organic chemistry, chemical engineering, and analytical science all to play a role.

in a national emergency, this time must be reduced by a factor of four, which will require a new culture of interdisciplinary research, collaboration among companies, and government intervention.[6]

A subtle but important consideration in the development of technology to meet chemical and biological threats is the different requirements for homeland versus military security. The tools and methods are similar, but the deployment strategies might be very different. Research in the chemical sciences must address these differences.

This workshop is one of six organized by the Board on Chemical Sciences and Technology of the National Research Council. The report is organized under three major headings: Grand Challenges, Specific Challenges (barriers), and Research Opportunities. Participants formed breakout groups to discuss each of these topics, and the next three sections summarize their discussions and conclusions. Discussions focused on how the chemistry and chemical engineering communities could assist with biological and chemical national security issues; although radiological issues were not a focal point of the workshop, the organizing committee recognizes the importance of this topic. The findings of the organizing committee are presented at the end of the report.

---

[6]M. Futran. 2002, Challenges in Rapid Scale-up of Synthetic Pharmaceutical Processes. Presentation, Workshop on National Security and Homeland Defense, Irvine, CA. (See Appendix D.)

2

Grand Challenges

Advances in chemical science and technology have enhanced national security in innumerable ways, contributing to our strength as individuals, our strength as a society, and our strength as a country. Chemical sciences and technologies produce most of the incredible range of materials that enable our modern society including clean water, abundant food, vaccines, and medicines; the clothes we wear; the homes, buildings, and factories in which we live and work; the roads, bridges, vehicles, fuels, and trade that support us; the electronic marvels that entertain, enable, and protect us; and the weapons and munitions that defend us. Chemical science and technology has markedly contributed to making us the envy of the world.

However, that envy also is a threat. Our success highlights economic disparity that may engender resentment and foment ideological conflict. Our chemical enterprise itself may become a target to be disrupted or diverted for retaliatory purposes by others. Discussions at the workshop have identified four grand challenges, the improvement and development of knowledge and technologies in the areas of threat reduction, preparation, situational awareness, and threat neutralization and remediation. Details are provided in this chapter and those that follow.

THREAT REDUCTION

Our economic success revolves around energy. Although we are only 4 percent of the world's population, we consume just under one-quarter of all the energy produced (and in so doing are the largest generator of carbon dioxide emissions from fossil fuel burning), but with that energy, we create more than one-quarter of the world's entire gross national product. Half of our energy comes from oil, and over half of that oil is imported. A quarter of our energy comes from natural gas, and an increasing percentage of that is also imported. Protecting our access to foreign sources of energy has a huge impact on foreign policy and national security.[1]

Eliminating our dependence on outside sources of energy would reduce perhaps the greatest threat to our national security. The United States has significant deposits of coal and uranium. It has significant reserves of oil and gas that are somewhat more expensive to extract than reserves from many foreign sources. It also has significant land area on which to place various instruments of energy capture such as windmills, solar cells, and biomass. The critical challenge to U.S. energy independence is the better exploitation of indigenous sources of energy, including solar energy. This challenge includes development of clean ways to use abundant coal; enhanced recovery of more of the oil and gas (and gas hydrates) we have; systems for the production, storage, and use of clean hydrogen as a transportation fuel; much less expensive photovoltaic, wind, and biomass systems for solar energy exploitation; safe and proliferation-resistant nuclear power and waste stewardship systems; and systems throughout for the minimization and more efficient use of whatever energy is consumed.

Energy independence was recognized by the workshop participants as one of the most important issues related to national security and homeland defense. As "Energy and Transportation" was the topic of another Challenges for the Chemical Sciences in the 21st Century workshop, further discussion of these issues will be curtailed in this report, but can be found in the report of the aforementioned workshop.

Many of the chemical products and services on which our society has come to depend would have a particularly strong or immediate health, economic, or military impact if they were to become even temporarily unavailable. The supply chains (raw materials, intermediates, catalysts, additives, etc.) essential for manufacturing these strategic products or services may be vulnerable to interruption, particularly if raw materials or intermediates are produced in few or off-shore locations, involve dangerous or unstable materials that cannot or should not be inventoried in large quantities, or are transported by vulnerable methods along vulnerable routes. A critical challenge to reducing the potential for chemical systems interruption includes the development of mitigation strategies, for example, the development of alternative sources of supply, establishment of strategic reserves, contingencies for rapid replacement of production capability, alternative transportation modes, and the development of alternative intrinsically more secure and less vulnerable chemistries.

---

[1]<http://www.eia.doe.gov/>.

The products of our chemical processing industries themselves could become the instruments of terrorists because of their flammability, reactivity, toxicity, or notoriety. It is critical to minimize the vulnerability of chemicals or chemical assets to attack, contamination, or diversion for terrorist purposes, particularly as weapons of mass destruction. Critical challenges include the development of systems or chemistries that reduce the amount of or substitute for materials currently at risk, alter the attractiveness of such materials to terrorists, minimize the inventory and transportation of such materials, and that can detect and track the covert production and transportation of such materials.

PREPARATION

We can only assume that terrorist attacks against people, the environment, commercial assets, and military assets will occur in the future. A critical challenge is to have integrated response measures in place to minimize loss of life and property from such attacks. Attacks employing chemical, biological, explosive, or radiological agents require different measures for response. For example, a chemical or radiological attack will present a hazardous materials problem because the chemical agents remain localized and the threat is not contagious, whereas an attack with biological agents will require a public health response because of the likelihood of spread beyond the point of exposure due to high communicability.[2] Integrated information systems will be required to coordinate and maximize the effectiveness of a diverse group of responders and a broad-based public health infrastructure must exist. Panic in the population will be reduced by education of the public about the nature of terrorist threats and how to respond to reduce injury.

To increase the effectiveness of our response to terrorist attacks, we need to understand the fundamental science of threat agents, including how, on a molecular level, they elicit their physiological responses in humans. Such fundamental knowledge could lead to the design of new antidotes and the development of effective prophylactic measures such as safer vaccines. We now face the possibility of attack by new biological agents produced through biotechnology. As a result, we require new technologies to rapidly identify novel agents and a capability to discover and produce medical countermeasures on an emergency basis. New technologies are needed in hospital emergency rooms to respond to mass casualties. New methods of medical surveillance could allow exposure of victims to threat agents to be detected and response to be initiated and coordinated prior to victims showing conventional symptoms.

---

[2]D. R. Franz. 2002, Current Thought on Bioterrorism: The Threat, Preparedness, and Response. Presentation, Workshop on National Security and Homeland Defense, Irvine, CA. (See Appendix D.)

SITUATIONAL AWARENESS

Loss of life or property from a terrorist attack at home or on our troops on a battlefield would be minimized or possibly prevented by the capability to detect the presence of chemical, biological, or radiological agents anywhere in the world. Detection capabilities are also required at borders and U.S. ports of entry, where shipping containers and vehicles should be checked for evidence of chemical weapons, biological weapons, and explosives. Beyond detection, the capability to unambiguously identify observed threat agents is required to maximize the effectiveness of our response.

To meet this critical challenge, new technologies for detection and identification of threat agents should be developed. The availability of detectors that are small, rugged, and low cost would allow our detection capability to be dispersed widely at home and abroad. The objective must be instantaneous detection and rapid identification, both with near 100 percent reliability, because erroneous

---

Industrial Use of Sensors
Scott D. Cunningham
DuPont

Most people do not think of the chemical industry as experts in sensor technology. Nevertheless, the industry is the largest designer, user, and consumer of sensors for risk avoidance and quality control. One group in DuPont identifies rugged and reliable sensors and analyzers for real-time process control. It is a difficult task to find a useful sensor in the real world. Take Raman or some of the fancier near infrared spectroscopies, for example; they are great tools in the laboratory, but to use them successfully on-line, in an oily environment, with dust and cleaning fluids residues flying around everywhere is a remarkable feat. In spite of this, these types of spectroscopies are running on-line in a nylon plant. In addition, fiberoptic-coupled sensors are used for on-line particle color, appearance, and size characterization, and sensors monitor the head space over fairly dangerous environments such as pesticides and hydrofluoric acid. These detection technologies can be and have been integrated into the everyday decision making and risk analysis used in real-world plant environments. The chemical industry has never dealt with sensing and detection in an office building or on the battlefield, and right now does not have an intuitive feel for what should be measured or how sensors would be installed. However, industry's talents in analytical chemistry, integration, and risk management can certainly be applied to mitigating the effect of terror in the nonindustrial realm.

detection or identification of threat agents can undermine the credibility and use of this capability. To achieve high reliability, sensitivity should be improved over what is provided by our current spectroscopic, mass spectrometric, and biological detection capabilities. Unambiguous identification of threat agents requires integration of detector response with databases of signatures of known threat agents, and the capability to recognize new threat agents from a generic property. The principle of orthogonality—multiple, uncorrelated analytical techniques in one sensor—must be included in the research and development.

THREAT NEUTRALIZATION AND REMEDIATION

If an attack on a chemical asset or by a chemical or biological agent does occur, it is essential that the damage be contained, neutralized, and remediated as expeditiously and safely as possible. Most chemical production and storage facilities already have disaster recovery and containment plans in place in case of an accident. These plans are also generally applicable to intentional attack. By design, damage to production facilities and the potential for massive chemical release, especially outside the plant boundary, is expected to be limited. An industry-wide system is also in place for response to transportation accidents,

---

Decontaminants
Mark Tucker
Sandia National Laboratories

Because decontamination occurs after buildings have been evacuated and first responders have treated any casualties, it allows for some time lapse before decontamination begins. With some planning, damage can be minimized and the efficacy of the decontamination process can be maximized. For example, sensitive equipment, electronics, valuable artwork, and personal objects will require chemicals that are less harsh than those used to clean air ducts and walls. The lack of time sensitivity also allows the choice of a decontaminant that may have a longer reaction time, but that is more suitable for the surface or the ambient conditions such as humidity and temperature. Thus, it is best to obtain a suite of decontamination methods.

There is currently a wide variety of decontaminants from which to choose, but many have not been tested or proven efficacious. Standard test protocols need to be developed to give regulatory approval to sensors and decontamination formulations and to ensure that a product is meeting the specifications of the users.

including fire suppression, leak containment, and chemical neutralization. Of course, assessment of industrial safety needs and installation of safety measures should continue.

Outside the chemical industry, however, organized systems are not in place to react to an intentional release of toxic chemicals or biological agents. Critical challenges for chemists and chemical engineers in this area include development of appropriate collective and personal protection systems for first responders, hazardous materials (HAZMAT) response teams, and the affected public; immediate and extended medical countermeasures for both chemical and biological agents; and neutralization, detoxification, decontamination, and disposal procedures for materials (for example, buildings and equipment) and environmental spaces (for example, soils and waterways).

3

Specific Challenges

In Chapter 2, a number of grand challenges for chemists and chemical engineers were delineated to promote national security and homeland defense, based on the views of workshop participants. This chapter goes one step further to identify scientific barriers that prevent those challenges from being met.

THREAT REDUCTION

Threat reduction involves any action taken prior to a terrorist act that serves to decrease the likelihood of an attack or to mitigate or eliminate the adverse effects of an attack on U.S. territory, on its population, and on its soldiers and property abroad. The best method of threat reduction would be to reduce and remove the incentives for terrorist acts, but this is beyond the scope of our charge. As a chemically related alternative, it would be desirable to have a detection system that can alert us to the mere possibility of attack and that may even avert a planned action. This system would be able to detect chemical and biological residues, and chemical and biological agents in sealed containers. Such a chemical detector could test for residues of weapons or their precursors on the hands of captured enemy soldiers, suspected criminals, suspicious dead bodies, or even airline passengers. With this type of technology, incidents could be prevented and weapons production laboratories large and small could be shut down, perhaps before production begins.

In the same vein, customs officials have stated that it is impossible to inspect all the containers entering the United States—in fact, currently only 2 percent of all containers are inspected.[1] An automated rapid-sensing detection system for chemical and biological weapons or precursors in closed containers would enable customs inspectors to ensure that a container's contents are indeed safe and to ensure the inspectors' personal safety if contents were a hazardous material. Automation also allows a much larger percentage of containers, if not all containers, to be inspected. This way, chemical and biological—perhaps even nuclear and radiological—weapons contents could be contained, neutralized, and controlled by the proper government authorities before entering the country.

Chemical and biological detection is a multifaceted puzzle. Some current detector technologies cannot achieve the sensitivity needed to detect trace amounts of chemicals that may be escaping from sealed containers or that remain on a suspect's hands. Some detection methods for chemicals and biologicals are not specific enough to sense agents through the background (smoke, perfume, or body odor, for example) collected concomitantly with the sample. Better specificity is also needed to sense agents when additional chemical or biological species are used that mask or mimic a weapon. Not all detection methods are quick, as needed for applications at sites like airports nor are all detection methods able to be used at varied distances. Although it is necessary to develop the ability to detect harmful chemicals through closed containers, suitcases, and other such items, it is also essential that the techniques are noninvasive and that the privacy of the individual is respected. Detection methods must also be nondestructive to the material from which a sample is taken. These are just a few of the representative barriers to improved detection technology.

Historically, the transportation industry has made strong efforts to advance accident prevention measures;[2] nevertheless, new and different actions may be needed to reduce the likelihood of a terrorist attack. For example, currently tank cars of chemicals that can react with each other to form a hazardous compound may be juxtaposed on trains. The chemical community could create guidelines, though the extra burden on the railroads due to the imposition of those guidelines may not be welcome and thus government regulation and enforcement may be required. Other barriers to safer transportation, such as appropriate placarding (labeling of tank car contents) or improved tracking systems, may be overcome by new regulations. For those regulations to be well-developed and scientifically sound, chemists, chemical engineers, and other scientists should assist in their development.

Chemicals in transport are frequently placed in storage tanks pending transfer to barges, ships, or tank trucks. Storage tanks are also used at manufacturing and

---

[1]R. L. Garwin. 2002, Thoughts and Questions on Countering the Terrorist Threat. Presentation, Workshop on National Security and Homeland Defense, Irvine, CA. (See Appendix D.)

[2]National Research Council. 1994. Ensuring Railroad Tank Car Safety. Washington, D.C.: National Academy Press, pp. 1, 4-7.

SafeT Pass
Richard L. Garwin
Council on Foreign Relations, New York

Approximately 1 million 8 ft x 8 ft containers enter the United States monthly, and only 2 percent of them are inspected in any way before delivery. These uninspected containers could hold nuclear weapons, contaminated items, biological weapons, or dissemination devices. How can these items be prevented from entering the country?

One safety measure discussed for trucks traveling from Canada is called SafeT Pass; it would work as follows. A truck carrying freight is inspected at its point of origin. Tags and seals are used to ensure that the consignment is safe and is not tampered with, that the truck has not had freight added along its route, and that the driver is certified and has been with the vehicle for the entire journey. SafeT Pass holders also have information on file showing that they have had an extensive interview and background check, which does not need to be repeated at each border crossing. This system could alleviate traffic problems at borders and ease the burden on inspectors.

---

industrial processing locations. Often, the storage areas are protected only by tall fences and perhaps security personnel. Storage tanks may contain tens of thousands of gallons of flammable or toxic chemicals that can be stolen or released with little effort. Thus far, storage areas and tank farms have not been troublesome, but the potential exists for terrorist actions using such tanks. Efforts must be made to make storage tanks more secure and especially to reduce the need for chemical storage by developing in-process chemical production and use.

Another barrier to threat reduction is public attitude toward chemistry, unawareness of the true nature of chemical and biological terrorism, and the risks involved in each. In an attack, damage inflicted by a chemical or biological weapon can be exacerbated by rash decision making and panic. Personal injury and even death could be minimized if first responders and the general public knew what steps to take to prepare themselves for entering and escaping contaminated areas, respectively. The immense nature of such an undertaking is overwhelming, and to find an organization to lead the effort would be difficult. Nevertheless, chemists and chemical engineers, together with other scientists, must play a large role in such a scenario.

Finally, the potential of threat is reduced if hazardous materials are less available. In this vein, more work is needed on process intensification and process redesign to minimize the accumulation and storage of hazardous intermediates and products.[3]

PREPARATION

The second grand challenge, preparation, is as important as threat reduction, since preventive measures will not preclude all terrorist attacks. Though chemical and biological weapons are many and varied, a common set of barriers in basic science exist that can be overcome to increase our readiness.

Medical Countermeasures

We need to have drugs, antidotes, and cures for the weapons of mass destruction that terrorists are likely to use. To develop medicinal countermeasures, the basic science involved must first be understood. For both chemical and biological weapons, the process of molecular recognition by elements of the human body is of utmost importance. However, we do not have a clear understanding of protein surface interactions, the relationship of genes to protein function, and how viruses infect and replicate. All of these processes are chemical in nature and cannot be solved without knowledge of the chemical sciences.

It is of great advantage to invest effort in pharmaceutical research before the need for certain drugs actually arises. As chemists, we can develop better methods for the identification of drug targets and drugs themselves, including cell-specific medicines. Antibiotic-resistant strains and genetically modified organisms are real concerns for which there is currently no easy solution. Theoretical and computational methods are not yet able to identify problems with therapeutic agents early in the design process. Testing protocols for broad spectrum antivirals should be developed and clinical trials of pharmaceuticals should be completed so the public can be confident in the aftermath of an attack.

Interdisciplinary and Multidisciplinary Research

The infrastructure of national security and homeland defense research needs improvement. Though many scientists are interested in assisting national security efforts, researchers who work with dangerous or infectious materials often have difficulties obtaining authorization for their work and even for receiving their samples. Additionally, research is not as efficient as possible due to lack of industry collaboration; of access to intellectual property (IP) and classified information; and of peer review of research proposals, engineered applications, and

---

[3]S. D. Cunningham. 2002, What Can the Industrial Chemical Community Contribute to the Nation's Security. Presentation, Workshop on National Security and Homeland Defense, Irvine, CA. (See Appendix D.)

A Lack of Vaccines?
David R. Franz
Southern Research Institute

To some, it seems that there has been a paucity of high-quality vaccines related to chemical and biological terrorist threats for U.S. residents. Until very recently, though, the Department of Defense (DOD) has been the only organization interested in vaccines for these types of threats. DOD has spent a large amount of money on the research and development for such vaccines with great success—in the last 8 years or so, approximately 12 new vaccines have been developed and proven effective on rhesus monkeys.

To advance development beyond this point becomes difficult, because DOD lacks knowledge and experience with pharmaceutical regulatory issues. They may need to partner with a large pharmaceutical company as a government contractor to develop, build, and run an agency similar to a national vaccine facility. Problems with this concept arise when indemnification is considered, because it is not only infrastructure that is of concern, it is also people. Vaccines invade the body, and the likelihood is that vaccines may cause two or three out of a million people to die. Neither the pharmaceutical companies nor the government is willing to take such a risk.

products intended for mass production. Some of these problems could be alleviated by creating a national center for chemical and biological testing, although such an endeavor would prove to be expensive.

Current intellectual property rules and regulations are hindering the progress of research that could be valuable to national security and homeland defense. One example, the inability to access internal IP, is a large barrier to development of basic scientific principles and related applied technology. Inexperience in technology transfer and the fragmentation of IP also makes it difficult for useful collaboration to occur. The National Technology Alliance, a government program, has succeeded in removing some of these barriers from the corporate world, but many more hurdles remain.

Graduate education and research is generally not designed to prepare students to address real-world issues in technology, especially national security related concerns. Students' fields of study tend to be narrow with limited exposure to research and issues in other scientific disciplines. The research that is envisioned as most beneficial to national security is often interdisciplinary—an intersection of chemistry, biology, physics, engineering, and materials science. Graduate education is slowly changing: there are some graduate departments that recognize subdisciplines such as biochemistry, organometallic chemistry, polymer chemistry, biomedical chemical engineering, materials chemical engineering, and environmental chemical engineering. Additionally, over the past 6 years, the National Science Foundation (NSF) has sponsored over 100 award sites through the Integrative Graduate Education and Research Traineeship program that promotes cultural changes in education by providing "a fertile environment for collaborative [multidisciplinary] research that transcends traditional disciplinary boundaries."[4] Nevertheless, graduate curricula do not usually reflect the interdisciplinary and multidisciplinary nature of research that would impact the technology base, economy, and safety of our society. To have the desired impact, changes are needed at the graduate level to properly prepare the next generation of scientists and engineers.

Related to interdisciplinary research and education is the notion of systems and the design and analysis of interrelated processes, with each individual process being subordinate to the system as a whole. Systems engineering considers how chemical processes are configured, controlled, and operated to maximize performance, productivity, safety, reliability, environmental impact, and other process attributes. It requires an understanding of basic chemistry and chemical engineering principles, but goes further by linking these principles to develop a chemical product or process. A systems perspective is rarely seen in graduate programs in the chemical sciences, but graduate students would be well served if they were at least introduced to such thinking because it would prepare them for the world encountered after graduate study. Certainly an appreciation of systems thinking would better prepare graduates of each discipline to contribute to the country's future needs in homeland security.

SITUATIONAL AWARENESS

The third grand challenge, situational awareness, poses the most numerous research challenges for chemists and chemical engineers. The science of sensing and detecting involves a wide variety of technologies and requires expertise in many different fields.

Sampling, though seemingly simple, is perhaps one of the most difficult steps in chemical and biological agent detection. The problems arise in ensuring that the sample taken represents the media sampled, which depends on proper local mixing. Concentrators must be developed that can circulate great quantities of air without clogging or losing effectiveness and that can concentrate the desired molecules. Recognizing and removing environmental interferences and background effects, for example, those due to geographical and temperature variability, cannot be done through improved sampling methods, but can only be accom-

---

[4]<http://www.nsf.gov/search97cgi/vtopic/>.

plished using a variety of techniques including electronic software, hardware, multivariate analysis, statistical analysis, and others. More efficient methods of separation, especially at the microlevel, are also crucial.

Preparation of an environmental sample for delivery to the sensor and the sample cleanup afterwards are often the rate-limiting steps in the detection of biological agents, as well. Even for biodetection, sample preparation is a chemistry and materials science issue, currently accomplished using membranes and surface-active chemistries, binders, and ligands. Biological sample preparation remains an embryonic field.

Sensors and detectors require different technology for different uses. Even the smallest tabletop mass spectrometer in a standard research laboratory cannot be converted into a rugged, lightweight model that a soldier, HAZMAT team member, or first responder could carry. Room temperature detectors; label-free detection (for example, without fluorescent beads); and low-cost, real-time analysis of airborne particles are barriers to widespread monitoring of chemical and biological weapons in public areas. Other obstacles that need to be surmounted by chemists and chemical engineers include improved nanoscale fabrication methods, better microfluidics and macro-microscale interfaces, better high-throughput screening, more efficient heat exchangers, and lighter batteries. Engineers are also challenged to integrate new technological components into a useful finished product.

To develop robust sensors, a multidisciplinary systems approach must be taken. Experimentalists, statisticians, engineers, and data analysts must be included from the beginning of the idea to the fielding of the sensor. Statistically designed experimentation must be used in the development of the field worthiness of a sensor, with statisticians a close collaborator of chemical and biological scientists. Actual and potential interferences must be identified and dealt with either through design of hardware, multiple sensor types, or multivariate techniques, or through software development such as statistical analysis. Sensor calibration and drift need to be addressed and corrected either mathematically or through chemical hardware.

Once developed, detectors and sensors need to be tested according to strict criteria. A well-defined and very demanding set of test standards, including limits of detection, has been in place for the last 15 years, initiated by the Department of Defense. The U.S. Army periodically offers the opportunity for realistic field testing, and wind tunnel testing is available at several official sites. It is imperative that scientists take advantage of existing resources for technology validation. It would be useful to develop new, integrated, multiple source databases to create chemical and biological agent libraries for quick agent identification and access to neutralization methods. Libraries already exist in many individual agencies, and there is a need for consolidation. Such libraries could also contain information on trace components of weapons that offer a signature of where or by whom the agents were produced to aid in identification of the criminals involved.

Detector Maintenance
Donald H. Stedman
University of Denver

The task of developing chemical or biological detectors for field use is not simple. Of equal importance, though, is the ease of maintenance of these detectors. During Desert Storm, the energy in the batteries of the M8A1 Automatic Chemical Agent Alarm carried by each soldier was continually low. The M8A1 works by pumping an air sample through a filter into the analyzer. Because the air in Iraq was full of sand, it clogged the filters very rapidly, which caused the air pumps to work harder and the instruments' batteries to die sooner than expected. The dead batteries would cause the alarm to sound.

Instead of replacing the air filters every three days as prescribed, it was, in reality, necessary to replace the filters every three hours. To perform maintenance every three hours is an unreasonable expectation— it is hard enough to do during the day, but it would also require a soldier to wake up multiple times in the middle of the night. Clearly, maintenance issues are equally critical to the usefulness of the instrument as is ensuring it works properly in the first place.

THREAT NEUTRALIZATION AND REMEDIATION

Neutralization and remediation of a chemical or biological threat is an area of research that has previously been confined to military and government laboratories. Now that such a threat is real, decontamination and remediation research must be expanded to academic and industrial laboratories. It has been made clear that civilian safety, too, is in jeopardy; hence scientists must become concerned with innovative and improved measures for collective protection (for example, filtration of chemical and biological agents from circulated air). Agent dispersal should be understood for scenarios at varied sites, various length scales (for example, human to room to region dimensions), and under different convective conditions due to weather or air circulation patterns. Also, there currently is no general-purpose decontamination method for all types of surfaces that can get into small crevices and is safe for use on all materials, nontoxic to humans, and environmentally benign. Standard testing criteria for decontaminants must be developed.

4

Research Directions

The grand challenges and specific challenges (barriers) posed in prior chapters should be met as quickly as possible to ensure the safety of the United States and its populace. Chemists and chemical engineers can help the nation by working together with other scientists to address questions relevant to national security and homeland defense in their own research programs if they recognize the need for multidisciplinary collaboration in research and in education.

THREAT REDUCTION

The actions that can be taken to reduce terrorist threats are many and varied. From increased energy independence to chemical transportation safety, from new methods of detection to intrinsically secure chemistry in production and processing, much of the research required for improvements in threat reduction technologies is similar to research needed for the other grand challenges. Synthesis of new molecules, synthesis of new materials, new synthetic methods, and new methods of sample collection and preparation could ultimately yield more highly selective sensors; more robust detectors; lighter, more portable detectors; more efficient energy conversion or storage; and manufacturing that is cleaner and safer. This could decrease the probability of attack on a chemical processing plant or allow the detection of illegal trafficking of weapons from stockpiles, among other possibilities. Because this research is so closely related to research in other areas, detailed discussion of research directions for threat reduction is included in the other sections of this chapter.

PREPARATION

The workshop identified the area of preparation, or pre-response, in terms of the following question: What can be done prior to a terrorist event to improve our capability to limit the effects of a disaster? The answers to this question encompass an enormous range of activities relevant to the chemical sciences. These challenges have been grouped into two subcategories: process engineering and chemical synthesis. Several issues that can be broadly classified under "infrastructure" also emerged in our discussions. These are treated separately below.

Process Engineering

Successful scaling up of chemical or biochemical reactors for rapid production of decontaminating chemicals or pharmaceutical products such as vaccines and antibiotics requires comprehensive understanding of scaling effects on the performance of a reactor. Quantifying reactor conversion and product selectivity requires knowledge concerning reaction pathways and reaction kinetics for the chemical or biological syntheses and rate processes of momentum, heat, and mass transfer for the specific types of reactors employed. The scale-up of the reactor often also necessitates the scale-up of other manufacturing process units. Given the need for reducing time from discovery to manufacturing, research into process intensification (reducing the physical scale of unit operations by orders of magnitude) should be pursued.

For pharmaceutical processing, other important unit operations include powder dissolution, crystallization and precipitation, drying, milling, blending, granulation, agglomeration, deagglomeration, compaction, lubrication, fluidization, encapsulation, tableting, and coating. These subjects are sparsely taught and inadequately researched in academe. Drug manufacturing and drug delivery systems need to be effective so that user safety is ensured in terms of therapeutic dosage applied and dosage effectiveness. Therefore, process control for each of the unit operations as well as quality control involved in the scale-up of the drug manufacturing process is of great importance for a reliable manufacturing process.

For pharmaceutical products, extensive knowledge of particle technology is necessary for the optimal design, operation, and synthesis of drug manufacturing units. The scale-up of these units has traditionally largely relied on empirical approaches due to complex powder or fluid-powder flow mechanics involved in a vessel of complicated geometry. More mechanistic approaches based on, for example, the similarity rule of dimensionless groups, leads to a very limited applicability with operation conditions restricted to only physical operation. A mechanistic scaling law verified under commercial-scale operating conditions for powder and fluid-powder reaction systems to be used with confidence for commercial unit scale-up purposes is currently not available. Replicating the units to increase production is the common viable alternative.

The volume of chemical components used for full-scale production of pharmaceuticals or other products and processes can be surprisingly large. Maintaining large stockpiles of toxic or otherwise hazardous materials is dangerous and inefficient. With the added risk of sabotage or intentional contamination, such practices are not prudent. The development of synthesis and fabrication processes that minimize the use or storage of large amounts of chemicals is a clear challenge to the chemical industry.

Scaling up solids processing systems is often more challenging than scaling up fluid processing systems, because fundamental understanding of particle technology for processing cohesive powders for pharmaceuticals is needed. For example, in powder blending (a critical unit operation in the drug manufacturing process) formation phenomena for a binary mixture of powders in a simple rotating cylinder are not yet understood. The more challenging problem of predicting powder-mixing patterns in complex geometric blenders also needs to be addressed. Computational fluid dynamics in the simulation of fluid or solids flows in reactive and nonreactive environments in various process systems could provide a useful alternative to the rapid system scale-up strategy. Computational code development, however, requires experimental verification to ensure its reliability.

Of course, the pharmaceutical industry is not the only sector in which process engineering is needed to enhance national security and homeland defense. Also, other areas of process engineering such as model development, optimization schemes, on-line control, catalyst development, new separation methods, biomass conversion, and models using multiple length scales are equally important to include in research efforts.

Chemical Synthesis

Many drugs are produced by either chemical synthesis or biosynthetic processes. Recent advances in synthetic organic chemistry, catalysis, biotechnology, and combinatorial chemistry have made it possible to synthesize many chemicals that are not found in nature or have heretofore been difficult to produce. Current chemical drugs, such as antibiotics, used to combat infectious diseases are threatened by bacterial abilities to quickly mutate into a drug-resistant form. Concern also exists for purposefully genetically modified organisms used for terrorist attacks. Consequently, there is a need to constantly develop new chemical drugs for fighting infectious diseases caused by new biological agents. As we know more about human genomics, many new drugs, whether small-molecule chemicals or large proteins, can be developed to better target the diseases.

Rapid production of small-molecule drugs will require the development of new organic reactions that maximally increase chemical complexity and that are highly selective. Advances in automation and miniaturization will be required to expedite discovery of synthesis sequences for large-scale drug preparation.

Synthetic Organic Chemistry
Mauricio Futran
Bristol-Myers Squibb

Even though synthetic organic chemistry is a very old scientific pursuit, I believe that today most medications needed for homeland defense will be created by means of synthetic organic chemistry. This type of research and development is done by industrial organic chemists and chemical engineers, and there are many opportunities to involve academic scientists to increase the speed of development.

Discovery of new biotransformations and improvements in separations and reactor design are also required. Developing a scalable high-efficiency separation process for separating and purifying chemical isomers, such as chiral compounds, would be useful for many chemical syntheses. In addition, the development of new drugs is dependent on new technologies for fast screening and testing. Fundamental understanding of biosynthetic pathways in cells, the structure-function relationship of biological molecules, antibody-antigen interactions, signal transduction on the cell surface, and the mechanism of toxicity are important. The ability to generate a mass library of chemical compounds and screen them for their biological activities or functions also remains a challenge to industry.

Genomics, proteomics, rational drug design based on structure-functional information, and knowledge in metabolic pathways and immunology also require further research and development. Development of new bioprocesses with a high yield, high product specificity, and high production rate remains a challenge that requires an interdisciplinary approach and collaborative effort among chemical scientists and engineers, biologists, and other scientists.

Of course, offering a new drug to the public is dependent on more than the synthesis. Preclinical trials, clinical trials, and regulatory approval tend to be rate-limiting steps for bringing a drug to market. In an attack, all of these processes may need to be streamlined and expedited through the use of new technologies, in addition to the use of more efficient drug development.

Materials synthesis is a necessary component in the development of advanced technologies for national security and homeland defense. For instance, new composites, nanoscale molecules and compounds, and polymers are needed for tougher, explosion- or puncture-resistant materials that can be employed in buildings, garments, bridges, and other products and structures. Personal protective materials could be enhanced with new chemical adsorbents; filter materials, impermeable membranes, artificial sutures, and improved energetic materials for

Enzyme Production for Natural Products Biosynthesis
C. Richard Hutchinson
Kosan Biosciences

Natural products biosynthesis relies on enzymes to create the final product. Production of these enzymes is in itself quite a process. Traditionally, the microorganism with the engineered gene that produces the enzyme undergoes random mutagenesis. The mutagenesis products are then screened for organisms that produce increasingly larger amounts of enzymes.

In today's laboratories, enzymes are produced by a more efficient process. Engineered genes are removed from the original microorganism and through DNA cloning methods are placed into a host that already contains abundant substrates from which the enzyme is made. This process may produce enzymes that do not function as well as science desires or as well as the native enzyme, but they do function. The overwhelming benefit of this process is that it requires 6 months of work, compared to 10 years of mutation and screening by the traditional method.

munitions and rocket motors could be developed. Simultaneously, new methods for tagging, tracking, and sensing precursors and other agents for explosives and nuclear, chemical, or biological weapons should be developed. New types of electrode and electrolyte materials, including biomaterials, are needed for high energy density batteries and fuel cells.

The ability to respond effectively to an event will require first responders and HAZMAT teams to coordinate thousands of details. Development of new materials for advanced telecommunications and radar could greatly improve the current response standard. Materials that can lead to faster computers, higher-density storage, and more efficient telecommunications are vital. One example of a basic area of research that could have an impact on our ability to respond to a threat is wide bandgap semiconductors, used, for instance, in phased-array radars. The development of shipboard phased-array radar systems over the past few decades has provided the military with a very high degree of situational awareness with respect to airborne targets.

Infrastructure

The need for a central laboratory or test facility to validate new detection, detoxification, remediation, and medical countermeasure technologies was repeatedly articulated at the workshop. Possibly, this can be expressed as a need for a central clearinghouse and also a need for a test facility (or facilities).

The Soldier Biological and Chemical Defense Command (SBCCOM) is located at the Edgewood Arsenal (Aberdeen Proving Ground) in rural Maryland, north of Baltimore. The SBCCOM is the Army's principal research and development center for chemical and biological defense technology and has an extensive complement of skilled professionals in this field. This program could easily be expanded for civilian defense needs.

Testing facilities for chemical and biological materials are expensive and complicated. Some exist in the private domain, for example, Battelle Memorial Institute. The Army's Dugway Proving Ground, located in an isolated area of Utah, provides tens of thousands of square feet of biological and chemical testing laboratories (40,000 instrumented square feet for biotesting; 35,000 square feet for chemical testing including 40 separate certified chemistry laboratories; and two 30,000-cubic-foot testing chambers). These facilities are available for private use on a contract basis and could be put to work on the demands of a civilian program. Expansion, however, may be required if civilian defense needs grow and cannot be reasonably accommodated in existing facilities.

New and improved instrumentation is also a necessary component of infrastructure. Single-user and multiuser instrumentation is needed to support research at testing facilities; obsolete equipment also needs to be replaced.

SITUATIONAL AWARENESS

Research needs for situational awareness stems from a number of chemical disciplines and other sources as seen below.

Dispersion, Collection, and Concentration

Atmospheric chemists and other scientists who have focused on pollution and global climate change have much to contribute to national security and homeland defense. These scientists have developed computational methods to accurately model and in many cases predict the transport of pollutants and particles in air or water. These same tools can be used to predict the effect of release of a chemical agent in an urban area so that appropriate emergency response plans can be developed.

Generally considered the "gold standard" in rapidly identifying small molecules such as chemical warfare agents, mass spectrometric techniques have been developed by analytical chemists that allow the identification of larger biomolecules. This work should prove important in identifying biological warfare agents including proteins such as botulinum toxin and viruses. Recent work has also focused on concentrating and identifying bacterial pathogens such as anthrax spores from air and water based on protein biomarkers. Finally, the development of miniaturization technologies by engineers to allow the manufacture of low-power, compact (portable) devices to perform these and other types of analyses is already a robust area of research in the United States.

The number of particles that can cause infection is often very small; hence, the amount of agent used in an attack can also be relatively small. Agent dispersion and sample collection, concentration, and preparation are all crucial issues for detection of biological warfare agents, chemical warfare agents, and explosives threats. The environmental sampling community (environmental chemists, atmospheric chemists) has been addressing the problem of sample collection and concentration for some time, although the need for real-time collection and analysis, though important in the national security and homeland defense arena, has not been particularly urgent. The medical diagnostics community has been actively engaged in developing methodologies for rapidly collecting, processing, and identifying disease in samples from patients, and recent efforts have focused on mechanical devices that can be used to provide a more active means of sample concentration and presentation to a sensor.

For example, new microfluidics technologies allow us to install "plumbing" on a small (millimeter-scale) device. The ability to accurately control the flow of liquids in miniaturized systems using current chip fabrication technology has been key to the development of the "lab on a chip," a low-cost, portable package that can be used by first responders, emergency medical personnel, and HAZMAT teams to analyze very small samples very rapidly. To extend miniaturization to the sampling and concentrating of aerosols and airborne particles, advances are needed in flow and handling of small volumes of gases.

Temperature Control in Microfluidic Devices
Andrea W. Chow
Caliper Technologies

One method of amplifying DNA in a polymerase chain reaction is through temperature cycling. The double-stranded DNA is denatured between 90°C and 100°C, while the annealing and extension demands lower temperatures. Sufficient amplification of the DNA occurs only after many cycles of temperature change.

DNA amplification can be achieved very effectively on a chip through Joule heating, heating due to energy loss from electrical currents. In some microfluidic devices, there are electrodes in the reagent wells to promote the flow of ionic liquids. Although the energy loss from these electrodes is often viewed as problematic, by varying the current to the electrode the temperature of the microfluidic device can easily be controlled.

Our ability to detect, identify, and track a plume of a chemical or biological agent in air or water is not adequate, and the capability to track long-range transport and to better understand survivability of bacteria and viruses over long distances is in great need. We also need to more completely characterize the natural background of spores and other bioagents in the 1 to 10 mm (respirable) range as it varies from place to place, seasonally, and temporally. An efficient approach to collect, separate, concentrate, and process samples has not emerged, and remains the significant challenge in this area.

Real-Time Detection—Sensors and Diagnostics

Identification of a radiological, chemical, biological, or explosive threat is achieved with a sensor. Diagnosis of disease is also often achieved with some sort of sensor, in the form of a medical test. Sensors can be remote, such as a satellite-borne camera, or they can be localized, such as a smoke detector or home pregnancy test. The chemical sciences have contributed significantly to the development of accurate and sensitive sensors for a variety of medical, pollutant detection, and industrial monitoring applications, and this robust area of research will continue to contribute in the national security and homeland defense arena. The challenge in this area is to develop specific, sensitive, low-power, fast, and robust portable devices that will detect radiological, chemical, biological, or explosive threats in the environment and that will rapidly diagnose disease.

Chemical Detectors

Mass spectrometry and ion mobility spectrometry are well-established analytical techniques that are heavily used by the DOD. These devices are used as detectors of chemical agents and explosives; in fact, ion mobility spectrometry is currently being used in most U.S. airports for explosives detection. Mass spectrometry is simultaneously broadband and specific—molecular masses provide exquisitely specific identification of chemical agents, precursors, trace explosives, and such materials. Of course, established methods can and should always be improved.

Improved detectors for conventional chemical warfare agents such as blood, nerve, and blister agents are desired. In addition, new methods to detect explosives and toxic industrial chemicals are needed. In general, the detection problem can be broken into two parts: chemical identification and signal transduction. The chemical identification issue involves some specific or nonspecific chemical recognition element that displays a unique response to the analyte of interest. Three general schemes for providing chemical specificity or identification were discussed by workshop participants: methods based on spectroscopy, cross-reactive arrays, and specific molecular recognition.

Spectroscopic methods involve measurement of characteristic "fingerprint" regions of the vibrational, mass, or electronic spectrum of the analyte of interest. There are many good examples of diagnostic spectroscopic tools that reside on a laboratory benchtop; chemists and chemical engineers are challenged to develop tools to enable the miniaturization and portability of such devices. Although mass spectrometers are not considered to be spectroscopic devices, small mass spectrometers are offered as an example of the concept of miniaturization in the Appendix.[1]

A cross-reactive array, or "electronic nose," behaves like the mammalian olfactory system: the response from an array of hundreds or thousands of individual sensor elements is combined to form a response pattern characteristic of the analyte of interest.[2] Various implementations of this basic concept have been developed, using solvatochromic dyes or electronic polymers as detection elements, for example.[3-5] Work in this field has established how multiple redundant sensor elements can be used to improve the signal-to-noise ratio of such an experiment, making analyte identification easier and clearer, and similar highly parallel assays have been applied industrially for the identification of proteins and sequencing the human genome.[6,7] Although the usual implementation of an "electronic nose" targets chemicals in the gas phase, this technique is equally applicable to detection of species in water or from blood or urine samples.

Nature provides many examples of how the specific binding of two complementary molecules can be used to achieve molecular recognition, such as an antigen-antibody pair or the DNA duplex. Indeed, many of the workshop presenters showed transduction schemes that incorporated biological molecules. This approach has the advantage that the sensor designers can rely on two billion years of evolution to design their recognition elements; this biological scheme works particularly well for detection of biologically derived molecules. If there are no naturally occurring molecules to be used as sensors, researchers can resort to chemical synthesis. Several molecules have already been constructed to specifically sense toxins, biological hazards, and explosives.

[1]K. A. Prather. 2002, Overview of Real-Time Single Particle Mass Spectrometry Methods. Presentation, Workshop on National Security and Homeland Defense, Irvine, CA. (See Appendix D.)

[2]T. A. Dickinson, J. White, J. S. Kauer, and D. R. Walt. 1996. A Chemical-Detecting System Based on a Cross-Reactive Optical Sensor Array. Nature 382: 697-700.

[3]J. W. Gardner and P. N. Bartlett (eds.). 1992. Sensors and Sensory Systems for an Electronic Nose. NATO ASI Series: Applied Science 212, 327. Dordrecht: Kluwer Academic Publishers.

[4]M. S. Freund and N. S. Lewis. 1995. A Chemically Diverse Conducting Polymer-Based Electronic Nose. Proceedings of the National Academy of Sciences U.S.A. 92: 2652-2656.

[5]N. A. Rakow and K. S. Suslick. 2000. A Colorimetric Sensor Array for Odour Visualization. Nature 406: 710.

[6]J. A. Ferguson, T. C. Boles, C. P. Adams, and D. R. Walt. 1996. A Fiber Optic DNA Biosensor Microarray for the Analysis of Gene Expression. Nature Biotechnology 14: 1681-1684.

[7]A. Lueking, M. Horn, H. Eickhoff, K. Bussow, H. Lehrach, and G. Walter. 1999. Protein Microarrays for Gene Expression and Antibody Screening. Analytical Biochemistry 270: 103-111.

For any of the recognition schemes described above, a quantifiable signal must ultimately be generated. This process is referred to as signal transduction. Participants in the workshop identified a variety of optical or electrical methods of signal transduction using a wide range of physical, spectroscopic, and materials tools. This is also a robust area of research in the United States that should play a key role in chemical agent detection.

The challenges in this field rest on improving sensitivity and specificity, and reducing the power drain to allow the manufacture of palmtop, wristwatch-size, or even smaller sensors. Advances in microelectronics have enabled the fabrication of compact, portable, low-power devices, and advances in power sources, miniaturization techniques, nanofabrication tools, and fundamental materials chemistry should allow this trend to continue.

Biological Detectors

A variety of detection methodologies that were either inspired by or derived directly from biological systems were discussed at the workshop. These have been used to detect conventional chemical as well as biological agents. Biologically derived systems may use just a small component of a living system, such as an antibody, or an entire live cell to achieve detection. For example, detection assays based on single cells or arrays of cells are now being used for diagnostics, in rapid drug discovery applications, and to sense toxins in water. Single components of biological systems, such as an antibody or a biomimetic membrane, have been incorporated into nonbiological systems to provide sensors for biological or chemical toxins. Chemical biomarkers released by the host in response to invasion and infection could provide a target for antibody arrays, mass spectrometry, and other analytical techniques to diagnose infection. In spite of these advances, general research into the biochemistry of agents and the rapid identification of pathogenicity of agents is still needed. This is particularly important if we are to develop the ability to respond to new threats such as artificially bioengineered diseases and agents.

The challenges in detection of biological agents are similar to those facing chemical agent detection. However, biological toxins can be much more toxic on a per-molecule basis than chemical agents and in some cases can be transmitted and amplified from one individual to the next. In addition, it is very difficult to distinguish toxic biological agents from the harmless biological compounds ubiquitous in our environment or from naturally occurring toxic biological agents that are present. As with chemical agent detection, significant challenges in this field involve improving sensitivity, specificity, and power requirements of devices. There are also opportunities for reduction of solutions and fluidics to create more robust, unattended, biologically based sensors. This area could benefit substantially from efforts to improve the response time and portability of medical diagnostic equipment. Advances in the medical field to improve our capability to identify pathogenicity by class or function as opposed to specific gene sequence will improve our ability to respond to unknown or emerging threats.

Other Detection and Identification Issues

Early detection and identification of nuclear, biological, chemical, and explosive warfare agents is essential for the timely deployment of preventive and defensive measures. One of the first steps involves constant surveillance of transportation in high-security areas such as airports and national borders. This can be achieved by using noninvasive sensors alone or in conjunction with imaging techniques capable of detecting and identifying sealed containers with chemicals or explosive materials inside, or items such as knives or guns on the body. These techniques include the use of ultrasound, optical waves, and nuclear radiation that are based on the interactions of acoustic or electromagnetic waves with matter. Ultrasonic and optical sensors detect reflected (or transmitted-through-and-reflected) waves. For nuclear materials, the detection and monitoring is primarily based on (isotopic) neutron and gamma ray detectors. Long-range detection of nuclear material is also necessary to prevent illegal trafficking from known stockpiles around the world.

As mentioned earlier, chemicals and explosives are usually detected spectroscopically. Fiber-optic-based chemical sensors have application in the real-time tracking of rapidly changing chemical environments. These sensors provide rapid

---

The Difficulty of Water Supply Contamination
Rolf A. Deininger
University of Michigan

Although worries abound over contamination of the water supply, in reality, the task is quite difficult to accomplish. For example, a contaminant can be dumped into a reservoir, but studies show that the chemical does not mix throughout the entire body of water, even after many hours. There are multitudes of chemical and biological agents that can be used to contaminate the water supply, but all contaminants do not behave similarly. Not all contaminants are threats—some become unstable in water, while others require such large quantities to do harm that they could never be dumped without being noticed. Additionally, if a disinfectant residual is maintained in the water distribution system, that residual will react with the contaminant, and the populace will remain relatively safe. It will be extremely difficult for terrorists to successfully contaminate the water supply.

and reversible response to a variety of chemicals at trace concentrations. Flow injection analysis on a microelectromechanical system (MEMS) microlab platform provides high sensitivity and selectivity within a matter of hundreds of seconds from a small sample volume. Micromachined gas chromatography sensors aid in on-the-spot, real-time chemical sensing of toxic gases. Further variations of miniaturized devices might include a tongue-on-a-chip that can identify water pollutants through colorimetric chemical reactions.

The main mode of proliferation of biological warfare agents is through the use of aerosols. Ultraviolet and laser-induced fluorescence techniques are under development for the remote sensing and detection of man-made aerosol clouds containing biological agents. The challenge extends to the detection of such agents that contaminate water and food supply lines, particularly close to end-users. For contamination of the water supply, the distribution system is the most likely candidate for an attack. An effective contaminant will be tasteless, odorless, and colorless, otherwise consumers will recognize the problem themselves without additional detection technology.[8] New detection technologies are needed for the in situ monitoring of biological warfare agents inside "suicide attackers" to curb the spread of contagious diseases, including smallpox. Chemical biomarkers released by the host in response to invasion and infection could provide a target for antibody arrays, mass spectrometry, and other analytical techniques to diagnose infection. There is also a growing need to detect mutagenic strains of existing agents and to render them harmless. Detection of these agents involves real-time size and number classification followed by multiple analyses to identify the virulence of the organism.

Research efforts are underway in the government and private sector to develop bioMEMS devices for multiple purposes such as drug delivery, microsurgery, and implants. Lab-on-a-chip platforms carry out the reactions involved in the detection and identification of DNA and protein sequences from minute sample sizes of biological warfare agents. The detection is achieved on a CD-sized device by chemical analysis of DNA/protein fragments released from the cell. These portable devices are engineered by using a combination of micropumps and valves. Mass-scale production of these devices usually involves a combination of techniques such as photolithography and electroplating on one hand and surface machining, atomic force microscopy-indentation, differential etching, self-assembly, and X-ray lithography on the other.

Devices that can be used by an individual without medical supervision (for example, biosensors incorporated in a bandage) need to be developed to detect and identify human exposure to chemical and biological agents. Current methods used to counteract human exposure to pathogens involve needle-based drug

---

[8]R. I. Deininger. 2002, Vulnerability of Public Water Supplies. Presentation, Workshop on National Security and Homeland Defense, Irvine, CA. (See Appendix D.)

delivery by trained personnel. Self-administration of required dosage (tens of milligrams) could be achieved by improved aerosol-based drug delivery devices, thereby eliminating the need for intervention by medical personnel.

Prevention of human exposure to chemical and biological agents must be achieved. On a large scale, this can be accomplished by maintaining the integrity of indoor air quality. Applying heat and temperature to the air that is taken in can change the physical characteristics of carrier aerosols and can lead to their aggregation and consequent removal by standard particulate filtration devices. At a more individual level, a rapid breath or urine test for appropriate biomarkers would be useful to identify carriers of communicable diseases (for example, small pox) before they have the opportunity to infect multiple people through actions such as boarding an airplane.

Miniature devices that work under ambient conditions (rather than at cryogenic temperatures) need to be developed to improve their portability. A biological aerosol sentry system developed to analyze data from numerous sensors can be used to detect the release of biological agents at public gathering arenas and closed spaces such as subways. Systems monitoring water, fuel, and food supplies would be necessary to prevent mass human exposure to chemical and biological agents. It is equally imperative to identify new strains of biological agents that are resistive to current medications. Further aerosol science research is needed to understand the fundamental mechanism of aerosol agglomeration processes to develop more effective preventive measures. More fundamental study of the interactions of acoustic or electromagnetic waves with matter is needed.

Information on Biological and Chemical Threats

For many of the currently available antibody and nucleic acid-based detection systems, identification takes advantage of unique characteristics of the pathogenic organisms in order to distinguish harmful from harmless species. However, discussion at the workshop pointed to a lack of knowledge and understanding of many of the characteristics of pathogenic or toxic agents. Some unique characteristic of pathogens may allow their selective collection or concentration out of complex air, soil, or biological matrices. To encourage and enable innovations in this area, workshop participants discussed potential advantages of an extensive database on the properties of pathogens that could be readily available to researchers considering solutions to the problem. A central, unclassified, Web-based repository for all the properties of pathogens and toxins could be developed. Table 4.1 below shows an example of some of the properties that could be listed using anthrax as an example. The table is broken into properties of the agent itself and properties of any of the known biological targets.

The database would present all known protein sequences and structures; nucleic acid sequences; exosporium structure; metric parameters such as mean density and size; spectral properties such as the fluorescence, fluorescence

TABLE 4.1 Sample Entry in a Chemical and Biological Agent Database

Anthrax Spores	**Biochemical Properties** DNA sequence Whole organism Specific toxin plasmids Proteins sequence and structure Exosporium Spore internal Toxins
	Physical Properties* Density average and standard deviation Size average and standard deviation Fluorescence Fluorescence excitation spectrum Scattered light spectrum Absorption spectrum (from microwave to UV) Raman spectrum Electrical conductivity, impedance Acoustic properties
	*Note some of these properties may be highly variable and dependent on specifics of sample preparation.

excitation, scattered light, absorption (from microwave to UV), and Raman spectra; and electrical and acoustic signatures. This database would focus on properties and would not contain information on the preparation or delivery mechanisms of agents. Thus, the information contained may not need to be sensitive or classified. Much of the biological data already exists on the Web and could be linked to the biowarfare database. For example, The Institute for Genomic Research maintains an on-line database of known bacterial DNA sequences.[9] Limited databases on these properties are already in existence, although they are not readily available. In addition, extensive information about and knowledge of chemical and biological agents is maintained at the Chemical and Biological Information Analysis Center (CBIAC) at Aberdeen Proving

---

[9]<http://www.tigr.org/tdb/mdb/mdb.html>.

Ground in Edgewood, Maryland. "The CBIAC generates, acquires, processes, analyzes, and disseminates CB [Chemical and Biological] Science and Technology Information. . . ."[10] The agency also maintains a hotline on epidemiological and medical treatment data for first responders. Maintaining a Web-based database would lower the barrier for researchers who may be thinking of proposing new and novel ideas for detection of biological and chemical agents.

THREAT NEUTRALIZATION AND REMEDIATION

In the case of exposure to chemical and biological agents, mitigation and neutralization efforts need to be immediate. The effective decontamination process requires decontamination techniques and equipment that are readily available or can be easily deployed. Prior to the actual application, it is necessary to secure proof of the decontamination process's efficacy by testing live agents, to demonstrate its versatility for a variety of surface structures, and to ensure that it is environmentally benign (possessing nontoxic and noncorrosive properties). It is clear that the decontamination agents need to be sensitive to the material surface, whether it is building material or human skin. For selection or development of decontamination equipment, it is necessary to consider the capability and throughput, effectiveness, set-up time, power capability, durability, and operational conditions. The decontamination processes could include physical processes (using solvents, sorbents, or filtration operation), chemical processes (reactive chemicals) and thermal processes (vaporization, for instance). Decontamination of human skin requires immediate action after exposure, using chemical, biological, or physical means with attendance to safety throughout the application. Information on common chemical agents and biological agents and their respective decontamination equipment is available in the *Guide for the Selection of Chemical and Biological Decontamination Equipment for Emergency First Responders*, published by the National Institute of Justice.[11]

The challenges in decontamination include the logistics of providing materials in a timely manner, the development of effective multifunctional decontamination agents that are able to neutralize a variety of chemical or biological contaminants of diverse properties, and the development of low-toxicity decontamination agents that can be applied to such sensitive surfaces as human skin. More importantly, agreement by regulating agencies must be reached as to what defines "clean", and a protocol for determining whether a facility is "clean" needs to be clearly outlined. Agreement between civilian and defense agencies as to what is

---

[10]<http://www.cbiac.apgea.army.mil/about_us/general.html>.

[11]National Institute of Justice. 2001. Guide for the Selection of Chemical and Biological Decontamination Equipment for Emergency First Responders. Guide no. 103-00. See also <http://www.ojp.usdoj.gov/nij/pubs.htm>.

"clean" would speed the development of disinfectants and their testing. Also, a method for transferring information from classified programs to civilian programs regarding methods for decontamination, results from decontamination tests, and how to best test a decontaminant would be useful.

5

Findings

The Challenges for the Chemical Sciences in the 21st Century Workshop on National Security and Homeland Defense brought together chemical scientists and engineers, from academia, government, national laboratories, and industrial research laboratories, who provided a broad range of experiences and perspectives. Their presentations and discussions showed that chemists and chemical engineers have a set of skills that map well into solutions of national security and homeland defense problems, and showed science and technology solutions to be even more promising when chemists and engineers collaborate with their colleagues in other disciplines. The organizing committee has extracted two multifaceted findings that are generalizations of the workshop discussions.

1. Extensive research opportunities exist throughout the many facets of chemistry and chemical engineering that can assist national security and homeland defense efforts.

    a. Research in systems and analysis is required to apply new knowledge of basic science to developing tools and products for national security and homeland defense. Three major needs were identified at the workshop:

- **Detectors with the broadest possible range for chemical and biological agents.** New sensing and sampling technologies will yield improved threat detection and identification technologies. Specific detectors that provide speed, high sensitivity, and great reliability are essential.
- **Miniaturization of detection and identification equipment for use in the field.** This is needed by soldiers, first responders, and other

emergency and military personnel. This research will also allow the development of detection systems on a chip. Understanding materials and transport phenomena at the micron scale may facilitate research on miniaturized processes.

- **Personal and collective protection measures.** Research is required to determine how to use new materials and methods to develop protective clothing, new drugs and vaccines, and infrastructure protection (air filters, stronger construction materials, decontaminants).

b. **Addressing research opportunities in manufacturing would better prepare the nation for a terrorist attack.**

- Minimization of the hazards of chemical transport and storage. Chemical transport routes by rail, truck, or barge pass through population centers and neighborhoods. Large storage tanks of chemicals often exist at industrial sites. The inherent danger of chemical transportation and storage can be reduced if new manufacturing processes are developed that use less reactive, less harsh chemicals or produce and consume the dangerous chemicals only on demand and on site.

- Rapid production scale-up procedures. If these procedures are established before they are needed, they can be implemented immediately in an attack. Although investigation into scale-up of all stages of production should occur, the scale-up of separation and purification is a specific challenge.

c. **Fundamental research in basic science and engineering is essential to meet these needs.** Much basic research stems from the need to solve specific problems faced by society. Supporting fundamental research for national security and homeland defense will enable discovery and understanding in many areas of chemistry and chemical engineering. These areas include materials development, pharmaceutical modes of action, particulate air transport and dispersal, catalysis, chemical and biochemical binding, solids flow and crystallization, energy sources and batteries, and sampling methodology and preparation. Research in many of these areas will be needed to meet the challenges outlined above.

2. Infrastructure enhancements are needed to promote the success of research in national security and homeland defense.

a. **Access to secure facilities.** Many important research efforts will require the use of hazardous materials, classified data, and specialized facilities. Collaboration with chemists and chemical engineers at government facilities will be necessary.

b. **Transformation of graduate education.** Changes in the infrastructure of graduate education to remove barriers to interdisciplinary and multidisciplinary research will enable students to expand their range of knowledge and to learn how to work in concert with others to achieve maximum effectiveness. This will be needed for them to participate successfully in any research program after graduation, especially one centered on national security and homeland defense related issues.

c. Instrumentation for research. Much of the needed research will depend strongly on new instrumentation to replace existing obsolete equipment as well as to allow new experiments. This includes both major multiuser equipment as well as specialized instrumentation for individual investigators.

Appendixes

A

Statement of Task

The National Security and Homeland Defense Workshop is one of six workshops held as part of "Challenges for the Chemical Sciences in the 21st Century." The workshop topics reflect areas of societal need—materials and manufacturing, energy and transportation, national security and homeland defense, public health, information and communications, and environment. The charge for each workshop was to address the four themes of discovery, interfaces, challenges, and infrastructure as they relate to the workshop topic:

- Discovery—major discoveries or advances in the chemical sciences during the past several decades.
- Interfaces—interfaces that exist between chemistry/chemical engineering and such areas as biology, environmental science, materials science, medicine, and physics.
- Challenges—the grand challenges that exist in the chemical sciences today.
- Infrastructure—infrastructure that will be required to allow the potential of future advances in the chemical sciences to be realized.

B

Biographies of Organizing Committee Members

John L. Anderson (Co-Chair) is a University Professor of Chemical Engineering and is affiliated with the Center for Complex Fluids Engineering at Carnegie Mellon University. He is also the dean of the College of Engineering. He received his B.S. from the University of Delaware and his Ph.D. from the University of Illinois. His research interests are membranes, colloidal science, electrophoresis and other electrokinetic phenomena, polymers at interfaces, and biomedical engineering. He is a former co-chair of the BCST and is a member of the National Academy of Engineering.

John I. Brauman (Co-Chair), the J. G. Jackson and C. J. Wood Professor of Chemistry at Stanford University, is a physical organic chemist whose research centers on structure and reactivity of organic and organometallic compounds in solution and in the gas phase. He received his B.S. in 1959 from the Massachusetts Institute of Technology and his Ph.D. in chemistry from the University of California at Berkeley in 1963. Brauman is a recipient of numerous awards, including the American Chemical Society's Award in Pure Chemistry, the Harrison Howe Award, and the James Flack Norris Award in Physical Organic Chemistry. He is chair of the Senior Editorial Board of SCIENCE Magazine and has served on several National Research Council panels and committees, including the Committee on Risk Assessment of Hazardous Air Pollutants. He is a member of the National Academy of Sciences.

Jacqueline K. Barton (Steering Committee liaison) is Arthur and Marian Hanisch Memorial Professor of Chemistry at the California Institute of Technology. She received her A.B. from Barnard College in 1974 and her Ph.D. from

Columbia University in 1979. She did subsequent postdoctoral work at both AT&T Bell Laboratories and Yale University. Barton's research areas are biophysical chemistry and inorganic chemistry. She has focused on studies of recognition and reaction of nucleic acids by transition metal complexes, and particularly upon DNA-mediated charge transport chemistry. She is a member of the Board of Directors of the Dow Chemical Company and the National Academy of Sciences.

Marvin H. Caruthers is professor of biochemistry and bioorganic chemistry at the University of Colorado. He received his Ph.D. from Northwestern University in 1968. His research interests include nucleic acid chemistry and biochemistry. His laboratory uses modern concepts in nucleic acid chemistry, biochemistry, and molecular biology to study regulation and control of gene expression. Caruthers is a former member of BCST and is a member of the National Academy of Sciences.

Liang-Shih Fan is Distinguished University Professor and chair of the Chemical Engineering Department at Ohio State University. He received his B.S. from National Taiwan University in 1970, his M.S. from West Virginia University in 1973, his Ph.D. from West Virginia University in 1975, and his M.S. in statistics from Kansas State University in 1978. He performs fundamental and applied research in fluidization and multiphase flow, particulate reaction engineering, and particle technology. Fan is a member of the National Academy of Engineering.

Larry E. Overman is Distinguished Professor of Chemistry at the University of California, Irvine. He received a B.A. from Earlham College and a Ph.D. in organic chemistry from University of Wisconsin, Madison. He is an organic chemist, specializing in new methods for organic synthesis, natural products synthesis, and medicinal chemistry. Overman is a member of the National Academy of Sciences and served as co-chair of the BCST from 1997 to 2000.

Michael J. Sailor is professor of chemistry and biochemistry at the University of California, San Diego. He received his B.S. from Harvey Mudd College in 1983 and his Ph.D. from Northwestern University in 1988. He completed postdoctoral work at Stanford University and California Institute of Technology from 1988 to 1990. His research focuses on the chemistry of nanophase semiconductors, phosphors, and biomaterials, with emphasis on chemical and biological sensors. His group has invented detectors for explosives and nerve warfare agents. He received the Arnold and Mabel Beckman Young Investigator Award in 1993 and was named an Alfred P. Sloan Research Fellow (1994-1995). Sailor received a National Science Foundation Young Investigator Award (1993-1998) and a Camille and Henry Dreyfus Teacher-Scholar award in 1994. He is a former member of the Defense Sciences Study Group.

Jeffrey J. Siirola (Steering Committee and BCST liaison) is a research fellow in the Chemical Process Research Laboratory at Eastman Chemical Company in Kingsport, Tennessee. He received his B.S. degree in chemical engineering from the University of Utah in 1967 and his Ph.D. in chemical engineering from the University of Wisconsin, Madison in 1970. His research centers on chemical processing, including chemical process synthesis, computer-aided conceptual process engineering, engineering design theory and methodology, chemical technology, assessment, resource conservation and recovery, artificial intelligence, non-numeric (symbolic) computer programming, and chemical engineering education. He is a member of the National Academy of Engineering and a former member of the BCST.

C

Workshop Agenda

**CHALLENGES FOR THE CHEMICAL SCIENCES IN
THE 21ST CENTURY**

WORKSHOP ON NATIONAL SECURITY AND HOMELAND DEFENSE

**Arnold and Mabel Beckman Center of the
National Academies of Sciences and Engineering
100 Academy, Irvine, CA
Monday-Wednesday, January 14-16, 2002**

AGENDA

Monday, January 14, 2002

7:30   **BREAKFAST**

SESSION I. CONTEXT AND OVERVIEW
8:00   Introductory remarks by organizers.
        Background of project.
8:00   **DOUGLAS J. RABER,** National Research Council
8:05   **RONALD BRESLOW AND MATTHEW V. TIRRELL,** Co-Chairs,
        Committee on Challenges for the Chemical Sciences in the 21st Century
8:20   **JOHN I. BRAUMAN**, Co-Chair, Organizing Committee for the
        Workshop on National Security and Homeland Defense

8:30   **DAVID R. FRANZ,** *Southern Research Institute*
        Current Thought on Bioterrorism: The Threat, Preparedness, and Response
9:05   DISCUSSION
9:25   **SCOTT D. CUNNINGHAM,** *DuPont*
        What Can the Industrial Chemical Community Contribute to the Nation's
        Security?
10:00   DISCUSSION
10:20   **BREAK**
10:50   **RICHARD L. GARWIN,** *IBM and Council on Foreign Relations, New York*
        Thoughts and Questions on Countering the Terrorist Threat
11:25   DISCUSSION
11:45   **LUNCH**

SESSION II. ANTICIPATION, DETECTION, AND RESPONSE
1:00   **ROLF I. DEININGER,** *The University of Michigan*
        Vulnerability of Public Water Supplies
1:30   DISCUSSION
1:50   **ANDREA W. CHOW,** *Caliper Technologies Corp.*
        Microfluidics: Development, Applications, and Future Challenges
2:20   DISCUSSION
2:40   BREAKOUT SESSION: DISCOVERY
        What advances or breakthroughs in the chemical sciences—related to
        national security and homeland defense—have been made in the past
        several decades?
3:45   **BREAK**
4:00   Reports from breakout sessions and discussion
5:00   **RECEPTION**
6:00   **BANQUET**
        Speaker—RALPH J. CICERONE, *University of California, Irvine*
        After September 11: An Expanded Agenda for Science and Scientists

Tuesday, January 15, 2002

7:30   **BREAKFAST**

SESSION III. REAL-TIME DETECTION
8:00   **DONALD H. STEDMAN**, *University of Denver*
        A Skeptical Analysis of Chemical and Biological Weapons Detection
        Schemes
8:30   DISCUSSION
8:50   **KIMBERLY A. PRATHER,** *University of California, San Diego*
        Overview of Real-Time Single Particle Mass Spectrometry Methods

9:20   DISCUSSION
9:40   BREAKOUT SESSION: CHALLENGES
        What are the grand challenges in the chemical sciences for which
        solutions would assist the nation's interests in national security and
        homeland defense?
10:45   **BREAK**
11:00   Reports from breakout sessions and discussion
12:00   **LUNCH**

SESSION IV. CLEANUP AND VERIFICATION

1:00   **MARK D. TUCKER**, *Sandia National Laboratories*
        New Approaches to Decontamination at DOE
1:30   DISCUSSION
1:50   **STEPHEN R. QUAKE**, *California Institute of Technology*
        How Integration Will Make Microfluidics Useful
2:20   DISCUSSION
2:40   BREAKOUT SESSION: TECHNICAL BARRIERS
        What are the technical impediments to solving the grand challenges?
3:45   **BREAK**
4:00   Reports from breakout sessions and discussion
5:00   ADJOURN FOR DAY

Wednesday, January 16, 2002

7:30   **BREAKFAST**

SESSION V. PRE-RESPONSE ACTIVITIES

8:00   **C. RICHARD HUTCHINSON**, *Kosan Biosciences*
        Biosynthetic Engineering of Polyketide Natural Products
8:30   DISCUSSION
8:50   **MAURICIO FUTRAN,** *Bristol-Myers Squibb*
        Challenges in Rapid Scale-up of Synthetic Pharmaceutical Processes
9:00   DISCUSSION
9:40   BREAKOUT SESSION: RESEARCH NEEDS
        What areas of fundamental research must be pursued to overcome the
        barriers?
10:45   **BREAK**
11:00   Reports from breakout sessions and discussion
12:00   Wrap-up and closing remarks
        JOHN L. ANDERSON, Co-Chair, Organizing Committee for the
        Workshop on National Security and Homeland Defense
12:15   **ADJOURN**

EXECUTIVE SESSION OF ORGANIZING COMMITTEE

12:15   Working lunch: General discussion
1:00   Develop consensus findings
1:45   Develop consensus recommendations
2:30   Develop action items, follow-up steps, and assignments
3:30   **ADJOURN**

D

Workshop Presentations

THE CURRENT THOUGHT ON BIOTERRORISM:
THE THREAT, PREPAREDNESS, AND RESPONSE

David R. Franz
Southern Research Institute

I have spent the last 15 years thinking about biological warfare. Unfortunately, over the last few months that topic has become extremely popular. Thus, this morning I am going to offer a broad perspective of biological terrorism, with the understanding that this biology is a subdivision of chemistry.

The last 60 years, as evidenced by the former U.S. and Soviet programs, has been the modern era of biological warfare. The agents studied in both nations' programs were very similar, and the workhorse agents in both programs were the zoonotic agents—agents that are transmissible from animals to humans (see Figure D.1). The physical and infectious properties of biological agents are not all the same: they differ in how they act; whether they cause illness or death; their stability during growth, production, weaponization, storage, and dissemination; the number of organisms that cause illness; and how infectious they are. The only seriously contagious biological agents, smallpox and plague, were actually weaponized by the Soviets and put into refrigerated nosecones of ballistic missiles to blanket the United States.

The fundamental differences between chemical and biological agents are very important as we look to the future. Chemical agents are volatile and dermally active and it is immediately apparent when there is human contact with a chemical agent. Biological agents, however, are not volatile, not dermally active, and

Human Diseases	Zoonoses	Animal Diseases
Smallpox	Anthrax	African Swine Fever
Cholera	Brucellosis	Foot and Mouth
Shigellosis	Coccidioidomycosis	Fowl Plague
	EEE / VEE / WEE	Newcastle
	Japanese B	Rinderpest
	Ebola/Marburg	
	Histoplasmosis	
	Melioidosis	
	Glanders	
	Plague	
	Psittacosis	
	Q Fever	
	Rabies	
	Tularemia	

FIGURE D.1 Of the diseases often associated with biological warfare, zoonotic diseases are the most likely to be used as weapons.

have delayed onset of disease. These properties place serious constraints on biological agent delivery. It is easiest to present biological weapons as irrespirable aerosols because they are nonvolatile, but it is a complicated task to prepare agents in that form. Difficulties also arise because distribution of the agent in air is completely dependent on meteorology (outdoors or indoors).

Biological agents also present problems to antiterrorist law enforcement officials. Although chemical weapons can cause fairly immediate death, biological weapons yield no sign of exposure. There are also no tools to detect exposure to a biological agent before the onset of sickness. Medical doctors are unfamiliar with exotic agents used in biological weapons, and the prevalence of flu-like symptoms in the beginning stages of agent-induced illness often leads to misdiagnosis. Not only does this yield a higher probability of serious health problems or even death if infected, but these factors feed an enormous psychological fear of biological agents. Other problems include the lack of "universal" vaccines, the lack of antiviral drugs, social or political issues revolving around prophylaxis, and the difficulty of forensics.

Biological terrorism is a unique threat because of its dual-use nature, evolving technologies, and political factors. The production of vaccines, for instance, requires the growth and subsequent death of viruses or bacteria. A certain veterinary vaccine facility in Russia is capable of producing 12 metric tons of foot and mouth disease virus in one run. It has been admitted that during World War II, the mission of this facility was to produce 240 metric tons of variola virus, which causes smallpox, to be used as a weapon. A similar dual-use situation exists with technology like crop dusters. It is difficult to know if pesticides are being sprayed or biological weapons are being tested. Biotechnology such as genetic modification adds yet another dimension to the prophylaxis/vaccine dilemma.

Political issues over the last 10 years have had a profound effect on the threat of biological terrorism. As the value of the Russian ruble fell and the country continued in its decline, between 30,000 and 40,000 Russian scientists and engineers, formerly employed at Ministry of Defense weapons facilities, lost their jobs. Perhaps these highly skilled scientists, who had families to feed and rent to pay, simply switched careers, but some were certainly recruited by Syria, Libya, Iran, and North Korea.

It is only in the last 4 or 5 years that the American public has become aware of the perceived terrorist threat. It is actually not significantly different from the threat during the Cold War. To produce an event that causes the death of thousands in this nation, the same agents selected during the Cold War must be used for the same reasons (ease of production, ease of distribution, rate of infection). Most likely, state sponsorship is required to produce the agents in the kind of formulation that would be effective for such a scenario. The exception is the use of a highly contagious agent like smallpox, which would not need to be weaponized.

Agricultural terrorism remains a possibility. The threat here is not to the human body; animal disease agents are not human pathogens and we could safely eat infected meat. The threat here is to the economy: in 1997 one infected and contagious piglet in Taiwan decimated the pork production capacity of the island, costing $5 million in initial damages and eventually costing $14 billion in lost revenue.

Where do the risks truly lie? The highest concern lies with highly contagious viruses like smallpox. Such a virus could cause the greatest damage; however, a smallpox attack is the least likely scenario to occur because the only legal supplies of the virus are in Atlanta and Novosibirsk. Illegal supplies probably do exist, but are difficult to obtain. Foreign animal disease viruses have the next greatest decimation potential. Classical agents that we normally think of but that are fairly difficult to deliver effectively are next on the list and then the hundreds of other more mundane agents like salmonella that will cause illness but not death. In the future, we may also need to beware of genetically engineered agents.

Bioterrorism may occur in the United States today because we cannot be beaten with conventional methods. We do not know how much "brain drain" occurred in the former Soviet Union, and the dual-use nature will never allow the problem to be dealt with through regulation. In the event of an attack, it would be ideal to have a universal detector that would simply indicate "yes" or "no" to the presence of any biological agent. The next steps would be to identify the tack, the agent, and the people involved, then to neutralize, decontaminate, and remedy the attack site. All of this would ideally be accomplished in 24 to 48 hours, and without any public panic.

Detection is an area that chemists have been involved in for quite a while. Detection of biological agents is much more difficult than originally believed at the beginning of the Gulf War. Although anthrax, with an infectious dose of 100 organisms per liter is relatively easy to detect, many agents like Q fever have an infectious dose of only 10 organisms, requiring sensors to detect 10 organisms in 100 liters.

Currently, biological agent detection in an infected person can only be accomplished by measuring the antibody response (unless methods like nasal swabbing are used, which have a high rate of false negatives, especially for agents that require a very small number of organisms for infection). However, it takes a few days for the antigen to circulate in the blood and for the immune system to respond (see Figure D.2). At this point, it is almost too late to treat the infected individual. Chemical and biochemical research needs to focus on tools to allow us to identify exposed individuals before the onset of clinical disease.

The question remains, how do current times compare with the Cold War era? It is now clear that there are people in the world willing to bring harm to civilians. Medical doctors must add new diseases to their differentials. The public now has

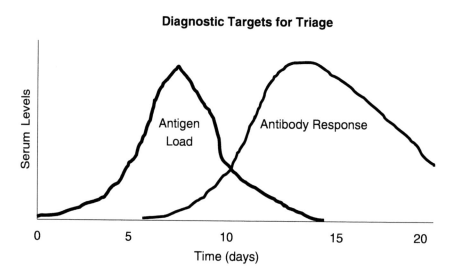

FIGURE D.2 Waiting to measure the antibody response to an infection allows the disease to progress too far.

a better understanding of the threat through the experience of the anthrax letters, and funding for counter-bioterrorism will increase. The nation's list of vulnerabilities is undeniably different. However, the technical difficulty of bioterrorism, the importance of meteorology, the difficulty of intelligence, detection and response (public health), deterrence and preparedness (law enforcement), and education on related issues have not changed at all since the end of the Cold War.

The solution comes as a two-pronged approach: education and a strong technical base. We have to educate people to be aware of the risk. Any developed nation in the world could produce biological weapons, and we, as scientists who speak the worldwide language of science, have an opportunity to prevent this. Our nation also must have a strong technical base to be ready for both the expected and the unexpected. This will improve our surveillance, diagnostics, and communication (data integration). So many unanswered questions remain regarding biological agents, the immune system, detection, and the like. The time is ripe for basic research.

The current bioterrorism situation is about more than just science and gadgets. I think it's reflected by the flags that I see on my street and on your street. It's reflected by the flags I see now on the lapels of business suits that I didn't see before. And I think it's about the American spirit: we're all Americans or scientists and we have a very important job to do.

WHAT CAN THE INDUSTRIAL CHEMICAL COMMUNITY CONTRIBUTE TO THE NATION'S SECURITY?

Scott D. Cunningham
DuPont

Today I am speaking on behalf of larger chemical companies. While I can only relate what I have seen at DuPont, my experiences are most likely typical.

The first response of the chemical industry to the events of September 11 was to donate materials needed in the recovery effort. Then, it began to think about its own vulnerability. Generally, the chemical industry enjoys a good safety record because of a necessarily heightened consciousness of the many hazardous materials handled. Recent events, however, have shown that technically knowledgeable people are willing to die in attacks against U.S. interests, and this is what causes concern in the industry. Companies began to rethink nearly everything, from infrastructure to checking identification at the door. They have changed capital expenditure plans, logistics, operations, and research and development. Industry members are now using the momentum of the September attacks to identify weak points in their systems and to reengineer operations. For example, storage tanks have became a major concern as targets of terrorist attacks, causing companies to redesign processes to minimize the accumulation and storage of hazardous intermediates and products.

As the resident HAZMAT authorities in many areas, chemical company employees found themselves responding to numerous community requests. There were so many calls for assistance that there was a shortage of trained responders to handle them. People began to think about communication devices, new systems, and new ways of doing things. Questions were asked like "What did I learn and how do I do better next time? When a call goes out, how do you mitigate the damage? What systems should be blocked off when entering a potentially contaminated building?" The experience gained through helping communities allowed chemical companies to improve their response procedures.

Chemical companies also received calls for many perceived antiterrorist materials such as filters, Tyvex, Nomex, Evacuate, and Kevlar, and pharmaceutical companies received calls for vaccines. The industry began to investigate who was buying these products and whether the products would meet the customers' needs. Materials scientists began to contemplate faster and different production methods, to combine products, and to define new uses for older products. Companies asked, "What do these things mean to the world? How can we make them better and protect our people and our food?" People responded creatively; for example, pharmaceutical companies are now working on faster vaccine production and thinking about new production paths through plants or bacteria.

In the corporate boardroom, the volition to act runs high, not to make a profit, but to help the national security effort. In fact, the chemical industry has a history of responding to national needs, evidenced by their production of black powder in the War of 1812, explosives and synthetic materials during the world wars, and materials for the space program. For this national effort, though, it is not clear whether there are technology solutions, and it is uncertain what the chemical industry can do to help.

The chemical industry has a proven track record of safety, and has helped companies to make their production plants safer. The industry also has global presence and experience, providing decision tree analysis and disaster planning around the world, including prevention, detection, damage mitigation, and mediation. The chemical industry is capable of integrating multiple sciences to produce a large amount of product at low cost. The industry has great experience with sensors, and is the largest designer, user, and consumer of sensors for risk avoidance and for quality control. These sensors have already been integrated into production plants, but not yet into office building monitoring.

Many chemical companies produce decontamination agents. DuPont, for instance, makes chlorine dioxide, the material used to decontaminate the Hart office building after anthrax contamination. DuPont did not perform the decontamination, however, because they did not know the effect of that product on a building, its infrastructure, computers, papers, or even the carpet. Although DuPont makes both the carpet and the decontaminant, we have never investigated the interaction of the two products. This is the kind of new, integrated thinking that is called for.

The chemical industry contacts supply chains, such as those for construction materials, automotive materials, and food, at multiple points. We can clean, disinfect, and genetically engineer seed. We can provide advanced coatings and packaging materials, as well as systems to detect bacterial contamination and spoilage. Can we make supply chains safer?

The industry wants to help the nation to get security systems where they are needed, at the right times and the right levels, but to do so industry must have more information about specific needs. The chemical industry brings expertise in product development, management, and integration. It wants to be useful. It wants to help mitigate risk. However, it has been very difficult for companies to get a clear idea of what the needs really are. Confusion about what is necessary does not inspire executives to spend employee time, money, and energy to solve a problem that has not been completely identified.

According to the Technical Support Working Group and the General Accounting Office, the chemical industry is focused on the short term. Certainly parts are. But perhaps 20 percent of research and development funding is focused on what the world may need next, things that are interesting but unsure. Yet there needs to be a better interface between what Defense Advanced Research Projects Agency (DARPA) produces and what industry does. There is a desire to spend more research and development money on terror mitigation, but there needs to be a better way to integrate the programs and systems that DARPA, the National

Science Foundation, and industry are funding, especially as the developer of each small component attempts to protect its own intellectual property.

The terrorist attacks have had a significant impact on the chemical industry. After looking to their own internal safety, companies want to contribute to the homeland defense effort. There is a lot to be gained from industry's scientific knowledge, culture of safety, role in society, science technology, and manufacturing. The chemical industry is willing to help, but obvious solutions are beyond any individual group or company. The process of formulating and coordinating contracts, objectives, and partnerships is underway, but it has only just begun. Getting all the groups and pieces lined up and together is going to take national will, a bigger sense of urgency, or some greater force. Perhaps organizations like the National Academies can help in this process.

THOUGHTS AND QUESTIONS ON COUNTERING
THE TERRORIST THREAT

Richard L. Garwin
Council on Foreign Relations, New York

As an resident of IBM's Watson Research Center, and especially as a senior fellow of the Council on Foreign Relations, New York, I'd like to share with you some of the national security and homeland defense related thoughts that I've had since September 11, 2001.

Unfortunately, terrorism and counterterrorism are enormous subjects that are not only vitally important, but are also urgent. A terrorist attack can come in many forms; the most worrisome of these is airborne bioterrorism. Anthrax distributed upwind of a city is a serious threat that could kill 100,000 people or more. More frightening, however, is an agent like smallpox that is not only infective, but contagious. Past experience with smallpox has proven its efficiency at causing tens of millions of deaths around the world.

Another tremendous threat is nuclear explosives, either stolen or improvised. A modern nuclear weapon could kill many millions of people. Even an improvised weapon in New York, on the level of the first-generation weapon used in World War II, could kill a million people. Additionally, there would be complete destruction of a 10-square-kilometer region and hundreds of thousands of people would be exposed to lethal fallout within the first hour. The devastation, however, is limited in extent, unlike a bioterrorism event, placing nuclear weapons second on the list of worrisome scenarios.

Bioattacks on food production need no explanation after seeing the results of the outbreak of foot and mouth disease in England. Since foot and mouth disease does not affect humans, the effects were primarily economic. Attacks on agricultural products would likely be just as damaging.

Based on the sarin attacks in the Japanese subway, chemical agent attack on humans has shown to be rather difficult. Biological toxins, chemical agents that are often overlooked, may also be difficult to use, but if used correctly can wreak much more havoc. Inhaling a single microgram of botulinum toxin is deadly, compared to a lethal dose by ingestion of 1 milligram for sarin and other agents. Widespread devastation may occur as a result of explosive attack on chemical plants or chemicals in transit, as seen in the accidents in Italy and Bhopal, India.

Also on the list of potential methods of terrorist attack is radiological attack. In the long term, relatively few additional people would die of cancer; psychological damage would occur in the short term.

Finally, the threat of a calculated explosive attack on structures still looms large after the three airplanes hit the World Trade Center and the Pentagon.

Through the Council on Foreign Relations, other scientists and I met with New York Governor Pataki's public safety officials to discuss specific counterterrorism measures for New York state. We discovered that links between the scientific community and state officials barely exist despite the knowledge that this type of communication is essential.

Collective protection using filtered air was one topic of discussion at the New York meetings. In the event of an anthrax attack, spores will drift into a building or house even with tightly closed windows. This problem can be alleviated with high-efficiency particulate air filters (HEPA filters) that have been in use since the Manhattan project. These filters are currently used in a number of places such as hospitals. In some buildings, HEPA filters can be easily interchanged for regular filters and their cost is not prohibitive.

A change of inside air for outside air needs to occur once every half hour. If the change occurs every few minutes for the circulating air, it would reduce the amount of biological agent that reaches a person inside a building by a factor of 10. If a HEPA filter is used in place of a normal filter, the agent dose is reduced by another factor of 10. Positive pressure within the building that allows no unfiltered air to leak in from the outside also greatly improves the protection level.

Countering explosives was another topic of discussion in New York. We need to be able to detect the presence of hundreds of grams of explosives on aircraft passengers and in bags. The simple act of bag matching is not sufficient, because today's terrorists will take their explosive-loaded bag onto an aircraft for a suicide mission. Explosives in vehicles, which can cause enormous problems if detonated at a choke point in the highway system or near a building, also need to be detected. Luckily, to do damage in this type of scenario, a large amount of explosive material is needed, which is hard to secrete in an ordinary car.

These are some of the issues currently facing us and that are being discussed around the nation. Scientists and engineers can provide options and can contribute to a rational decision-making process helping to make our nation safer for our residents.

VULNERABILITY OF PUBLIC WATER SUPPLIES

Rolf A. Deininger
The University of Michigan

Water supply systems are vulnerable to destruction and contamination. This is nothing new and has its roots in antiquity. A late director of the FBI called this to everyone's attention,[1] and his prescription for defensive measures were gates, guards, and guns. He also recognized the threat by a disgruntled insider and advised a background check on all employees. In 1970, the World Health Organization (WHO) published a booklet on "Health Aspects of Chemical and Biological Weapons." One appendix of that publication deals with the sabotage of water supplies and discusses agents, scenarios, and expected outcomes of a contamination. There will be a new release of this booklet in 2002.

The major elements of a water supply system are the raw water source (lake, river, reservoir, or groundwater aquifer), the water treatment plant, the pumping stations, and the distribution system consisting of pipes and intermediate storage reservoirs.

Dams on rivers can be destroyed by explosives, not only leading to a loss of the source, but possibly causing serious damage and loss of life due to flooding. Pumping stations are not easily replaced and may require considerable time for repair since replacement pumps are not usually stored on site and may require a lengthy ordering time. A partial destruction is not too serious since there is usually spare capacity allowing a somewhat reduced service. Destruction of pipelines is easily accomplished, but can be fixed in a short time since such occurrences normally happen in distribution systems due to earthquakes and other natural disasters. Utilities are well prepared for such events. There is a great deal of redundancy in a distribution system, and several key elements may be taken out of service without removing the ability of the system to provide service to the consumers at a reduced rate. The key to a higher level of security is therefore the redundancy of the system. This redundancy should be investigated for each particular system to ensure security compliance.

A water supply system may be contaminated at the raw water source, at the water treatment plant, or in the distribution system. Contamination of the raw water source is easily accomplished since it is usually at a location far from the service area. It is not, however, a very effective measure due to the large quantity of water involved. A contamination at the water treatment plant would be more effective, but increases the likelihood of detection since a water treatment plant is staffed around the clock. In addition, the treatment processes will reduce the contaminants by one or two orders of magnitude. Thus the distribution system is

[1]J. E. Hoover. 1941. Water Supply Facilities and National Defense. Journal of the American Water Works Association 33(11): 1861-1865.

Suggested Reading on Water Safety

1. Berger, B. B. and A. H. Stevenson. 1955. Feasibility of biological warfare against public water supplies. J. AWWA. 47(2): 101-110.

2. Clark, R. M. and R. A. Deininger. 2000. Protecting the Nation's Critical Infrastructure: The Vulnerability of U.S. Water Supply Systems. Journal of Contingencies and Crisis Management. 8(2): 73-80.

3. Delineon, G. P. 2001. The Who, What, Why and How of Counter Terrorism Issues. J. AWWA. 93(5): 78-85.

4. New York Times, Dec. 31, 2001. Files Found: A Computer in Kabul Yields a Chilling Array of al Qaeda Memos.

5. World Health Organization. 1970. Health Aspects of Chemical and Biological Weapons. Annex 5. Sabotage of Water Supplies. pp. 113-120.

the most likely candidate for an attack, especially the distant locations. For this scenario the technology of treatment at the water plant is irrelevant. For a contaminant to be effective it must be tasteless, odorless, and colorless. If these criteria are not met, consumers will recognize a problem and avoid water usage.

There are many agents that can cause serious health consequences or death when introduced into a water system. This has been summarized well by Burrows.[2] A rough classification of the agents might be chemical warfare agents, biological agents (protozoa, bacteria, viruses), toxins, and the large group of toxic industrial chemicals (TICs). Chemical warfare agents are normally deployed through the aerosol route and contamination of the water is a secondary effect. They are not a credible threat, and can be removed in the treatment processes.[3]

The most dangerous agents are the biological agents and the toxins. Protozoa, for example *Cryptosporidium parvum*, have contaminated a water supply system with serious consequences. In 1993, the supply of Milwaukee was contaminated, 100 people died, and over 400,000 became ill. Although this protozoan causes serious health effects in the very young, the very old, and the

---

[2]W. D. Burrows and S. E. Renner. 1999. Biological Warfare Agents as Threats to Potable Water. Environmental Health Perspectives 107(12): 975-984.

[3]National Research Council. 1995. Guidelines for Chemical Warfare Agents in Military Field Drinking Water. Washington, D.C.: National Academy Press.

immunocompromised population, it is not a credible threat to the majority of the population. Among the most dangerous biological agents are *Bacillus anthracis, Shigella, Vibrio cholerae, Salmonella,* and *Yersinia pestis.* For these agents, infectious doses are not very well known, nor is their survival in distribution systems that carry a disinfectant residual. Data are available only from mice and primates, and are usually expressed as an LD_{50} (lethal dose where 50 percent of the exposed die). There are serious questions whether this is a proper criterion, and perhaps an LD_{10} or even lower percentage is appropriate.

The previously mentioned 1970 WHO publication sets the calculations for contaminating a water supply. It assumes an infectious dose of 10^6 organisms in a glass of water (200 mL). Consider a medium-size city reservoir of 10 million gallons (40 million liters) or about 200 million glasses. The number of organisms necessary would be 200×10^{12}. Freeze-dried bacteria have about 10^{11} bacteria per gram. So 2,000 g are sufficient, if they can be mixed into the water, to contaminate the entire reservoir. On the other hand, the lethal dose of botulinus toxin is about 1 microgram. Thus 200 g (1/2 lb) are necessary to contaminate the reservoir such that each glass would be lethal. These are extremely dangerous agents and the only line of defense is to maintain a chlorine residual in the distribution system.

TICs are threats to water systems, but are not as problematic as chemical or biological agents. Consider cyanide with a lethal dose of 25 mg. To contaminate the reservoir mentioned above would require $25 \times 200 \times 10^6$ mg, or roughly 5,000 kg (5 tons). Dimethyl mercury is lethal at 400 mg; more than 50 tons would be required. Of course, it is possible to contaminate a small reservoir or simultaneously a group of several reservoirs.

We are not defenseless. There are many small steps that can be taken to harden water facilities against an attack. More timely monitoring of the water quality is imperative. At the moment, monitoring of surrogate parameters is possible, but we need to develop on-line direct measurement techniques of pathogens and toxins.

MICROFLUIDICS: DEVELOPMENT, APPLICATIONS, AND FUTURE CHALLENGES

Andrea W. Chow
Caliper Technologies Corporation

The recent development of microfluidics technology has mainly been driven by the need to miniaturize, integrate, and automate biochemical analyses to increase speed and throughput and reduce costs. These advantages can provide significant benefits to many applications for national security and homeland defense, including rapid detection and analysis of biological and chemical warfare agents and facilitating the development of therapeutics against these agents. Although the technology is still in a relatively early stage of development, a few commercial microfluidics products are now available and are beginning to demonstrate the technology's inherent advantages. This paper will review a number of current applications of microfluidics and outline some future challenges for new applications for use in national security and homeland defense.

In many microfluidic devices, a microchannel network is microfabricated onto a substrate (glass, quartz, or plastics, for example), and the channel plate is bonded to another piece of substrate with access wells matching the ends of the microchannels. This allows buffers and reagents to be supplied through the wells (see Figure D.3). For controlling the flow of fluids and reagents in the microchannels, electrokinetics and pressure-driven flows are the two most common means used. These driving forces can be applied to the reagent reservoirs through the instrument interface to the chip with no active fluidic control element fabricated on the chip.[4] In addition, a number of laboratories are developing active pumps and valves on chip for flow control.[5] Using these driving forces with on-chip or off-chip fluid control, it is possible to emulate many functions on chip, including flow valves, dispensers, mixers, reactors, and separation process units. For reactions such as polymerase chain reaction (PCR) that require elevated temperatures, precision temperature control of the fluid in a microchannel has also been demonstrated.[6]

One important area of technology development is the world-to-chip interface. The requirements for an effective interface are ease of fabrication, low dead volume, ease of automation, no sample biasing from the reagent source, and compatibility with existing sample storage formats. One such interface for high

[4]A. R. Kopf-Sill, A. W. Chow, L. Bousee, and C. B. Cohen. 2001. Creating a Lab-on-a-Chip with Microfluidic Technologies. Integrated Microfabriated Biodevices, M. J. Heller and A. Guttman, eds. New York: Marcel Dekker.

[5]P. Gravesen, J. Branebjerg, and O. S. Jensen. 1993. Microfluidics—A Review. Journal of Micromechanics and Microengineering 3: 168-182.

[6]M. Kopp et al. 1998. Chemical Amplification: Continuous-Flow PCR on a Chip. Science 280: 1046-1048.

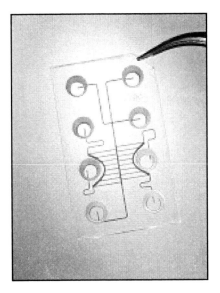

FIGURE D.3 Caliper's microfluidics chip fabricated in glass.

throughput screening uses technology that has a low dead-volume fluidic coupling of a capillary to a microchannel to enable automated sampling from microtiter plates, a standard format used by pharmaceutical companies for compound storage and bioassay analysis.

Microfluidics-based products for research and high throughput experimentation now exist in the commercial market. Applications such as genomic analysis, clinical diagnostics, and home-care products are undergoing various stages of development as well. For the research instrumentation market, the Agilent 2100 instrument and LabChip™ DNA sizing application were the first microfluidics products, introduced in 1999.[7] In the DNA sizing application, gel electrophoresis is performed in the microchip for up to 12 samples placed into the reagent wells. The samples are serially injected into the gel-filled separation channel, and a fluorescent intercalation dye is used to stain the DNA fragments. As the fragments move down the separation channel by electrophoresis, a laser placed at the end of the channel excites the fluorescence of the intercalated dye, and a photodiode detector measures the fluorescence intensity of each DNA fragment as a function of transit time. Using calibration markers to co-elute with each sample, the DNA size and weight fraction of each fragment in the sample can be determined accurately and automatically with the analysis software.

---

[7]<http://www.chem.agilent.com/Scripts/PDS.asp?1Page=51>.

For diagnostics, microfluidics has been useful for significantly shortening the analysis time for molecular fingerprinting. The combination of Bacteria BarCodes' rep-PCR technology with Caliper's LabChip™ technology has demonstrated in a feasibility study to provide a much quicker turnaround time than conventional slab gel electrophoresis for bacterial strain identification in samples. The rep-PCR technology is based on the discovery that repetitive sequences are interspersed throughout the DNA of all bacteria studied to date. The spacing between these repetitive sequences varies among bacterial isolates because their DNA is different. After DNA is extracted from bacterial cells, reagents for PCR are added, along with *Taq* DNA polymerase and primers that bind to a repetitive sequence. Since repetitive sequences are highly conserved in all bacteria studied, one set of universal primers can be used on almost every isolate.

During the amplification cycles of the PCR, the *Taq* DNA polymerase copies the DNA between the primers. Because numerous areas throughout the bacterial chromosome are amplified, many potential genetic differences can be detected. Figure D.4 shows a chip analysis of two samples of bacterial DNA amplified by rep-PCR. The gel electrophoresis results clearly show that the bacteria samples are different strains, as there are band differences appearing at approximately

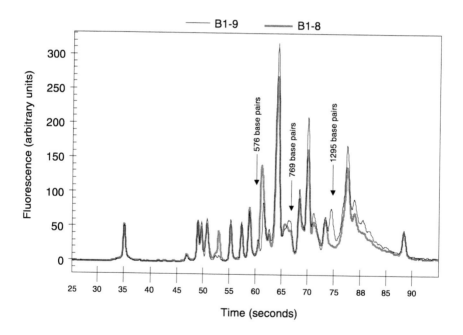

FIGURE D.4 LabChip™ DNA sizing data showing two different bacterial strains after rep-PCR of two samples, B1-8 and B1-9.

576, 769, and 1295 base pairs (marked by arrows). This is one example of an existing microfluidic application that can be used for rapid diagnostics following a bioterrorism attack.

Microfluidic technologies are likely to play an important role in high throughput screening of drugs and antidotes.[8] Enzymes and peptides can be mixed in a microfluidic reaction channel. Fluorescence intensity measurements can be used to determine whether the addition of a potentially pharmacologically interesting compound actually has an inhibitory effect on the enzyme. One advantage of using this technology over conventional methods is the reduction in reagent usage, which is especially important for proteins that are difficult to produce or purify. Other advantages are speed, higher data quality, and improved reproducibility due to the increased level of automation. Such technology has possible application for antidotes for chemical and biological weapons.

For other national security and homeland defense applications, specific requirements will no doubt impose new technological challenges and motivate new innovations in the field of microfluidics. Future challenges include developing suitable world-to-chip interfaces for sample collection (from air or liquid sources), devising methodologies to ensure statistically meaningful sampling (especially when the sample concentration is very low), integrating functions for on-chip sample preparation including sample stacking, understanding and mitigating undesirable surface interactions, developing generic detection methods for high sensitivity and specific detection of a wide range of analytes, and executing effective functional and system integration to yield robust and reliable microfluidics products.

[8]S. A. Sundberg. 2000. High-throughput and Ultra-high-throughput Screening: Solution- and Cell-Based Approaches. Analytical Biotechnology 11: 47-53.

AFTER SEPTEMBER 11: AN EXPANDED AGENDA FOR SCIENCE AND SCIENTISTS

Ralph J. Cicerone
University of California, Irvine

Tonight I want to give you some largely personal thoughts and observations. One message is that the real edge that the United States has in all of its challenges is science and technology. It's been that way for a long time, over 50 years now. I will argue that we have had previous large crises during which agendas changed, and we have risen to the challenges. Previously and once again, the United States' advantage is science and technology. There's great potential and great responsibility for the chemical sciences community because of that. Our focus is on chemical topics, but it could equally well be communication technology.

Let me go back into the past a little and talk about some other major shifts that have occurred, changes probably as big as the one since September 11. Let's go back to the World War II period, skip ahead to the interval after World War II to the end of the Cold War, and then discuss the period after the end of the Cold War. In our attitudes toward our government and the private sector and how science serves them, pronounced shifts occurred. To close, I will talk about the post-September 11 period and some questions that have arisen. None of us know exactly what all of our bearings will be, so I will address some of the questions that I think all of you have on your minds. The topics that you've been talking about at this workshop are a good representative sampling of all the things we have to think about, many of which we have not thought about before.

World War II

Until World War II, the size and scope of the United States' science research effort were both very small. Those of you who have read about Vannevar Bush will know part of this story. Vannevar Bush (an MIT electrical engineer) helped to convince the United States government in the early stages of the Second World War that the government should broaden its approach: instead of just issuing specific contracts to certain government-corporate labs or to a few universities for specific tasks, the United States should broaden the approach and take a more long-term view with more grant-like support to corporate and university labs.

The successes of science and technology were quite great, including the Manhattan project and the development of radar at MIT with some related work at Columbia University, developments which, along with their British counterparts and Allied fighting forces, led to victory in the Second World War.

Now, it's hard to realize just how little scientific research was being supported in the United States up until the middle of World War II, but it was minuscule. There were a few people who were convinced that basic and long-term research, not just that applied to specific projects, was worthwhile, and they held sway after the Second World War. The Office of Naval Research had become convinced to support pure mathematics even during the war, and to this day the Department of Defense maintains and supports some of the strongest basic research in the country.

After World War II and the Cold War

After World War II it was seen that the successes of science and technology in the war were so great and the threats of the emerging Cold War that were beginning to be felt in the late 1940s were so large as to encourage the country to create the National Science Foundation (NSF) and some national laboratories. New grant programs were spawned at the NSF, the Department of Energy (it was called the Atomic Energy Commission and other titles), and at the Department of Defense (previously the Department of War). Today's National Institutes of Health, I think, probably would not be what they are without the model of NSF. Also during this time, the National Aeronautics and Space Administration was created largely because of the threat that the Soviet Union's Sputnik represented.

Subsequently, basic scientific research made great contributions to our nation's economic strength and its security. Some wonderful examples are described in a series of publications called "Beyond Discovery: The Path from Research to Human Benefit" produced by the National Academies. Each brochure chronicles some discovery in basic science and how it has had applications beyond what anybody could have imagined. Often as a side benefit, many new products emerged from federal government support (for example, communications satellites) and a series of new discoveries was made.

One example is the development of the Global Positioning System (GPS), which came out of early basic work on atomic clocks and hydrogen masers and led to timing devices that no one could ever have anticipated. These timing devices have enabled positions to be measured with tremendous accuracy and precision using instruments on Earth-orbiting satellites, with applications from national defense to personal safety and convenience. Similarly, cochlear implants that enable hearing arose from early fundamental research in anatomy, electrophysiology, and information theory. Developments in laser physics have led to new types of eye surgery, the science of polymers has led to new fibers and artificial human skin tissue, and basic research in genetics has led to many beneficial applications in medicine and agriculture. In this post World War II– Cold War period, basic long-term research was understood to be good for the country to ensure immediate applications to specific defense projects.

In the Cold War there were also some specific needs that energized basic research. For example, prior to communications satellites, using the ionosphere to reflect radio waves was the only way to communicate around the world at the speed of light. In World War II, military powers experimented with the Luxembourg effect (heating the ionosphere) to jam communications and to prevent the opponent from communicating quickly around the curvature of the Earth. Such phenomena and their challenges generated a lot of research; in this case, some pretty good physical chemistry research had to be conducted to understand plasma in the upper atmosphere, in particular, ion-molecule reactions. Understanding of collective phenomena in physics was also gained in this same pursuit.

This kind of rapid and robust progress had not occurred before the Second World War. The key point is that research came to be seen as necessary for national security. A scientist who was very eminent in national policy told me that there has never been any real belief in the value of basic research for its own sake. Instead, it has been supported by the government (since the late 1940s) because of national security or more recently, to develop the economy. It was the need of national security that kept the research engine going.

More broadly, this new, widespread faith in science and technology and its successes on the military front led to the creation of something called the National Defense Education Act of 1958. The NDEA did not simply send people to graduate school in physics and chemistry and mathematics and engineering. Instead, there were NDEA graduate fellowships in fields like history, geography, and the foreign languages, and it provided support for science and math teachers in public schools. All of these good things were driven by national security needs. The perceived need for a strong defense capability and national security provided support for all of higher education.

Basic research in corporate labs also flourished after World War II. Just think of the fantastic capabilities and first-rate science that was associated with corporate labs like those of Bell Labs, IBM, Exxon central R & D, Eastman Kodak, Polaroid, Xerox, and others. Notice that I said "was associated," because those labs (that have survived at all) are now much smaller and are focused on shorter term projects. During the Cold War, some classified research and development went on in those labs also, and national security was a central priority. The phrase "good enough for the government" originated in the Second World War when factory workers and researchers had to produce something that was good enough for the government. In my lifetime, however, "good enough for the government" has been a put-down.

Confidence in government also characterized the time after World War II and during the Cold War. Over the history of our country it has been seen that one of the legitimate roles for the federal government is to provide for the national defense. There's been a lot of disagreement about many other tasks but not over this one. Military and security needs carried a lot of development throughout the post-World War II period. Of course, the Vietnam War led to less confidence in the government and caused the termination of classified research projects at universities.

1989 Until September 11, 2001

Now we come to the end of the Cold War and another large shift. Not only did science and technology carry the United States through the Second World War and the Cold War period stably and securely, but in the meantime scientific research and development were growing the U.S. economy. There are a number of economists (growth theory economists) who have quite a bit of evidence to show that more than half the economic growth in the United States since 1945 is due to scientific research and development (R & D). There is some argument among economists over how much is attributable to basic research, but if you lump together all research and development, well over half the growth in the U.S. economy is due to scientific R & D, and again, government sponsorship is right at the core of it.

Until the end of the Cold War there was a great deal of confidence and faith in government and in science because of its achievements such as its role underlying economic growth. Winning the Cold War was great, but it also turned out that nobody (that I know) saw it coming or knew what to do next. The end of the Cold War and the fall of the Berlin Wall in 1989 left everybody in a state of surprise. We didn't know what to anticipate.

Without the threat of a 10-foot-tall enemy who had launched Sputnik and had nuclear weapons just like ours and was out to destroy us, we turned away from science and higher education. The end of the Cold War undercut the support for science and for higher education in general. The National Defense Education Act fell into disrepair. Even scientific research, especially basic research, was under great threat and question. I would like to quote Roy Schwitters, who was the head of the Superconducting Supercollider (SSC) project in Texas. When Congress pulled the plug on the SSC, Roy was quoted in a newspaper as saying that the SSC would be just the first of all the big science projects after the end of the Cold War that would be subjected to very intense scrutiny. No longer would the undercurrent of trust and respect for science and higher education carry projects like that. I think Roy turned out to be correct. (Of course, there were other issues with the SSC.)

At about the same time, the end of the Cold War, Japan not only emerged as a high-tech economic competitor for the United States, but also as a winner. It was only 10 to 12 years ago that we thought we had lost the whole show to Japan in terms of R & D leading to high-tech and high-value products and the knowledge industry. Do you remember? And because we no longer had the large military and ideological enemy to fear, the public was asking science and the government, "What have you done for me lately? And by the way, you're too expensive." At about the same time, the globalization of the economy arrived and our corporations no longer had the kind of edge that they had had historically. Exposure to low labor cost all around the world undercut prices here and made it very difficult for our leading corporations to maintain robust central R & D labs that could do long-term research as well as product development. This was a big change, and the agenda for science changed again. Now, instead of providing for the national security and having faith in basic research to provide eventual long-term growth, there was less threat to the national security, and the need of corporations for commercializable products was immediate. Those powerful corporate research labs that used to do basic long-term research vanished.

The good news is that many of those corporations still do marvelous product development and they are holding their own in the world, but they do not do as much long-term research as they did pre-1989. In addition, the national laboratories that were created after World War II and during the Cold War were told to forget about weapons and to do something useful, to create products and do commercially relevant research. Thus the agenda for the national labs also changed. In fact, government's role in general was reduced. The private sector was supposed to take over. The private sector was featured and respected, and the government was less respected.

American corporations have done well post-1989, in the face of having new developments imitated within a period of months after great development costs. As in any new high-value product, the knock-off and imitation industry takes over. Our corporations can stay ahead of the game only due to the research and development that they can draw from. This is true in electronics, computers, memory devices, pharmaceuticals, aerospace, and so forth. But, once again the corporations cannot afford, ever again, because of the globalized economy, to maintain the central R & D labs that could focus on long-term basic research that was not product driven.

The diminution of long-term basic research in corporations opened the doors for universities to take on the role. The national labs were certainly capable of contributing, but they were not encouraged much to do so. At the same time universities were being given a different agenda, especially in science, and that was now not just to teach and train the next generation of people and to do research, but also to spur the local economy, the regional economy, and the national economy.

The 170 or so research universities around the country are under tremendous expectations to take on an additional role: to provide a real boost and a driving force for the economy. This is feasible because with our dual role of teaching and exposing students to up-to-date technologies in our labs they are ever more ready to go out into industry. And we can also do research on campuses without paying the wages of expensive researchers like the corporate R & D labs had to do, because graduate students and postdoctoral fellows do some of the research.

Did the universities actually rise to this challenge? I'm not sure. I don't know of much quantitative evidence, but generally I think you'd have to say basic research in the United States is still very strong.

Another characteristic of the post-Cold War period, 1989 to about now, but pre-September 11, is that corporate/university partnerships were supposed to flourish. We had laws enacted like the Bayh-Dole Act, whose purpose was to spawn the commercialization of university research, to permit university faculty members and researchers to patent things along with their universities even if they had been supported by the federal government. This is what Birch Bayh and Bob Dole provided for. This has been a success in terms of obtaining more patents and more commercialization.

At the same time, though, there was less support for government in general. And the expenditure of government funds on research was now being questioned, just like Roy Schwitter said it would be, and we have challenges like the Government Performance and Results Act, GPRA, which many of you have spent some time trying to understand. All federal agencies that support research now have to write a report every year and convince the Office of Management and Budget that they are being productive, they are accountable, there's no duplication, there's no waste, there's no repetition, and this is all helping the country in a productive way, preferably today.

This questioning came out of the era when there was less threat to national security, more emphasis on competing with Japan, and more emphasis on the private sector. People were asking, "How much research is enough?" People like Ralph Gomory (a mathematician now at the Sloan Foundation) tried to take that question seriously. Ralph and his colleagues through the Committee on Science, Engineering, and Public Policy (COSEPUP) proposed a reasonable goal for the United States: we must be excellent in some fields and while we cannot be excellent in all fields, we have to be good enough in every field to be able to recognize large breakthroughs. If we are really substandard in some field, then presumably we wouldn't be good enough to recognize big breakthroughs elsewhere. COSEPUP produced such assessments on mathematics and materials science. The point of this story is that there were real efforts and incentives to try to justify science on grounds other than the national security, usually competitiveness and commercialization.

Post-September 11, 2001

Since September 11, 2001, a lot has changed. Suddenly we are aware of new threats to our security. They're everywhere, and the topics of this workshop represent many of them. Water supplies, biological agents, exotic explosives, transport of airborne pathogens, security of chemical plants, use of nuclear waste for harm, civil structures, and so forth. Science and technology, however, are once again our advantage. And I'm very confident that the agenda of scientists and the agenda of the public and its attitudes toward science will change favorably if we do our work.

It is only through science and technology that we can leverage our otherwise small numbers of people. Science and technology must help us to anticipate, prevent, and/or mitigate destructive attacks.

How can we mobilize to be most effective? This workshop is a great start. I was very impressed when I heard Alice Gast give a Board on Chemical Sciences and Technology briefing back in October because, in fact, chemistry is the central science, and it will be in the nation's response.

What shifts are just beginning now and what is the expanded agenda for science and technology? The challenge to us as scientists is to capitalize on our available knowledge and to develop new knowledge and substances and instruments from it, and also to develop new knowledge where it's needed for the conventional military and for antiterrorism in general.

As we face this new and expanded agenda, we will all have to answer some questions and develop some new patterns. No one is sure exactly how this will happen, but one initial reaction after September 11 is to look to the government. That is what people are doing right now. Why? Well, who were the heroes on September 11 and since? They were the firefighters, policemen, policewomen, emergency workers, and the military, all public employees. So there is renewed confidence in the role of government. I don't know how long that's going to continue, however.

There was over-reliance on the private sector, from 1989 to September 11, 2001. One obvious example is the privatization of airport security operations that occurred in the previous 15 years.

While we don't know how this is going to play out, much of the necessary research and development that we're going to need in the coming years is unlikely to be generated by profit motives, at least initially. The initial development, inventorying, and coordination we have to do is unlikely to be taken over by a profit-making company. Perhaps the initial tendency to turn to the government will continue for a while. However, I believe that we need the strengths of the public and private sectors and that total reliance on one or the other is dangerous.

How do we inventory all relevant knowledge and coordinate the R & D? To start, the National Research Council (NRC) is really stepping in with the BCST at the front of it, and the newly reorganized Office of Science and Technology Policy is active, too.

This job is hard. Probably just today some of you are hearing about developments and interests from colleagues you didn't know you had who are doing things that you had no idea anybody was doing. That's the hard part of it.

But the good news is that the job is hard because our R & D enterprise is so vast. There is so much going on out there in university labs, national labs, and corporate labs, that just doing the inventory of all the potentially applicable science and technology that's out there is going to be difficult.

How do we coordinate this work to go forward from where we are when there are many disciplines involved and also many entities? That's going to be tough. Can we change the missions of the federal agencies rapidly enough and responsibly, for example, without giving up other essential parts of their missions? We need people in charge of the federal agencies and the state governments who are very quick and responsive. With the very small incentives we give for people to serve in government, this is a real handicap. So I think it's going to require a lot of volunteer work from companies and universities, public agencies, and the NRC to make it happen.

Who will do the work? This is a very serious question. As the years pass, and this post-September 11 condition persists, which I think it will, we will need more holders of graduate degrees, for example. One of the many threats out there is that it might become harder for foreign students to enroll in U.S. graduate programs or even to maintain their enrollment, depending on political attitudes. And right now the enrollments in these programs in the United States are mostly noncitizens, especially in engineering.

So my question is, who will do the work? How can we attract more U.S.-born students into graduate study in science and engineering if it becomes necessary, which it might? We really need ideas. Will it happen naturally? Some of us are in science and engineering because of our reaction to Sputnik and what our teachers told us in reaction to Sputnik. Reactions like that that can occur again.

There is also a strong need for people with advanced capabilities in foreign languages and deep understanding of history, other cultures, and some social sciences. The National Defense Education Act seems to have been needed even at the depths of the Cold War, 1958 to 1970, which I think is roughly when it ended. We should reinvent or reinvigorate the NDEA, to include graduate study in languages, history, international affairs, and similarly valuable fields.

Another question that may arise is, will classified research grow again, and what will be our attitudes toward it, for example, on university campuses? Faculty members who grew up in the 1960s may not want to see classified research again in universities. And there are fundamental issues, too. The obligation and necessity for faculty members to conduct free inquiry with the ability to publish freely is a major principle that we hold.

What about the old question, how much research is enough? Will it come back? How will we answer this question? I want to say strongly that when we try to coordinate our R & D effort, we should not start with the view that we should eliminate all overlap and repetition in science funding. A colleague who served on the National Science Board in the mid-1970s came to believe that overlap and repetition is needed to assure that all the necessary research gets done.

So when we talk about federal agencies starting to coordinate with each other, let's all speak up and say that effective coordination does not mean the elimination of all repetition and overlap. To do so would guarantee that all the research that's necessary does not get done. The current reality, however, is that all federal agencies are still coping with GPRA. How can we remind everyone of the need for basic research even when its potential applications are not evident now? Other forces also continue at play that make the job even harder. Research and development must continue to provide economic competitiveness and profits and satisfy the immediate needs for devices and substances to assure the safety and stability of society.

There are many, many examples of the fruits of basic research, and I think the NAS publication series "Beyond Discovery" is something that we have to get out into the hands of journalists and the general public so that they understand. We have to make these facts known much more widely, that the results of basic research from NSF, DOE, and NIH are very potent and important.

To conclude, I don't know how to answer these questions, but I do think that we can strongly assert that science and technology are our advantage against terrorism. More narrowly, I think the results in Afghanistan to date demonstrate the role of science and technology quite clearly, and I hope that people are catching on. We don't know yet how to use all of our resources here at home, but we do know that science and technology have a lot to offer, and chemistry in particular will once again be central.

We can also be proud of the entire scientific endeavor. The scientific enterprise in this country, indeed the world at large, represents some of the best attributes of human civilization. Nobel Laureate Robert Wilson spoke around 1969 about basic research in the Fermi lab: "This new knowledge has all to do with honor and country, but it has nothing to do directly with defending our country except to make it worth defending."

A SKEPTICAL ANALYSIS OF CHEMICAL AND BIOLOGICAL WEAPONS DETECTION SCHEMES

Donald H. Stedman
University of Denver

One joy of being around for a long time is that you can become awfully skeptical. You see things that are supposed to be new, wonderful, and just about to happen, when suddenly a vast amount of money is given to someone else, you see the results, and you say, "I could have spent one-tenth of the money for ten times the result."

There have been many quite useful discoveries in chemistry and chemical engineering over the years that have been used for detection applications. The first example is plasma chromatography, otherwise known as ion mobility spectrometry. In the 1980s this technique became the method of choice for detecting chemical warfare agents and was used by soldiers in Desert Storm with unfortunate results. Official reports tell that the rate of false alarms for these instruments was so high that soldiers became desensitized to real hazards. One infantry battalion eventually turned their alarms off. Much of the Gulf War Syndrome may well have been caused because ion mobility spectrometry was oversold as a detection technique.

Fourier-transform infrared spectroscopy (FTIR) is a very good tool that has been used successfully both in laboratories and the field. With enough optical path length, chemicals can be detected at parts per billion levels in only a few minutes with good resolution. Long-path FTIR has been used successfully in Utah where chemical weapons are destroyed. Nevertheless, FTIR has been oversold as a long-range, look-ahead detection tool. At White Sands, the Defense Threat Reduction Agency dropped large bombs on bunkers containing chemical agent simulants to determine the extent to which the agents were destroyed or dispersed. FTIR was used to measure the dispersed agents. Amazingly, passive FTIR would only work if the sky was blue and the ground warm, providing temperature contrast, or if readily detected SF_6 were added to the simulant.

Mass spectrometry (MS) is highly selective. The ability to further perform tandem mass spectrometry (MS/MS) analysis when a compound is detected to confirm the detection virtually eliminates false positive and negative alarms. But MS/MS analysis must be completely automated for the average GI to be able to perform it. A clever hand-held chemical and biological mass spectrometer has been developed that weighs only 4.3 pounds. The problem with the unit is production of the necessary vacuum, which requires 35 amps at 24 volts. Thus, battery-operated portable mass spectrometry is not yet available.

Chromatographic methods have included development of element-specific atomic emission, flame photometric, and flame chemiluminescent detectors. For example, a flame chemiluminescent phosphorus detector has been suggested for chemical weapon detection and does work for simulants. However, since ordinary flame photometry is sufficiently sensitive, it makes little sense to measure chemiluminescence, which requires hydrogen for the flame, a vacuum pump on the tube, and an ozone generator among other equipment.

Mike Sailor of the University of California, San Diego, has recently developed an element-specific fluorine detector to be used as a portable nerve gas sensor. What makes his instrument so different is that it has been presented to the scientific community (at an American Chemical Society meeting) before it is put into the hands of soldiers. This gives the opportunity for peer review and for corrections to the technology, if needed, to ensure that the instrument is useful and that money isn't wasted or lives aren't endangered.

The National Institute of Justice has put together multivolume compendiums of instrumentation relevant to chemical and biological weapons detection. However, none of these books contains a critical review of the effectiveness of the technologies. One instrument included in the publication is a portable, hand-held, ion mobility spectrometry chemical agent monitor with moderate to high selectivity, but only when used in open spaces, far from vapor sources such as smoke, cleaning compounds, and fumes. This would seem to make it useless in the battlefield. Another listed chemical agent monitor has a "below 5% false positive rate." With one in 20 false positives, no one could reasonably act upon an alarm.

There are many barriers to and challenges for instrumentation commercialization. It seems that most often, the parties involved are working at cross-purposes. The (often academic) inventor is never satisfied and always wants to continue to improve his or her invention. The instrument company only wants to build and sell the product. They would also like to have planned obsolescence to guarantee future sales. The salesperson has to sell what he has now, while convincing the customer that it will meet all his needs. The customer wants a product that is inexpensive and will last forever.

In the meantime, congress wants the military to solve the chemical and biological detection dilemma and military field commanders do, in fact, need the detection problem to be solved. Although the military can write technical specifications for detection equipment, the contract still goes to the lowest bidder who may not actually be able to deliver the specified product. The soldier is left with what is given him, whether or not it is actually a working, useful tool.

Perhaps scientists need to educate the public that it is impossible to have a perfect instrument. If zero false negatives are required for a detector, then there must be some amount of false positive responses. This does not mean that science has failed: risk is inherent in everything.

Conversely, the risks taken by the soldiers that now have Gulf War Syndrome were unnecessary. Beta testing of new technologies should have been done and should still be done before those new technologies enter the field. Because of classification and other problems, there is no peer review, leaving no one to stop disasters before they happen. While an individual scientist advising the military that the $30 million they just spent on an instrument was wasted might be shooting himself in the foot, a large organization like the National Academies or a large chemical company might more easily and truthfully review an expensive military product without repercussions.

The skepticism of the chemical community can truly provide a service to the nation.

OVERVIEW OF REAL-TIME SINGLE PARTICLE MASS SPECTROMETRY METHODS

Kimberly A. Prather
University of California, San Diego

Today I am going to speak about some of the continuous aerosol mass spectrometry methods that are currently in use. These methods are primarily used for atmospheric chemistry measurements related to human health such as pollution remediation, although now national security and homeland defense applications are starting to evolve. Most aerosol analysis methods are not real-time analyses; bridging the gap between on-line and off-line technologies is a challenge that is being addressed.

Typically, sample collection for aerosol analysis is accomplished by pulling particle-containing air through a filter. Collection times for obtaining enough sample to analyze are typically long. This fact precludes the observation of short-term variations, and also makes it possible for sample integrity to be lost during the long period between collection and analysis. How representative the collected sample is of actual atmospheric compounds is an additional question.

Mass spectrometry is an excellent analytical tool because it has extremely high sensitivity. With time-of-flight mass spectrometry (TOF-MS), if the ions are focused properly, it is theoretically possible to obtain up to 100 percent detection efficiency. A wide range of analytes can be detected, from very small organic or inorganic molecules to proteins and macromolecules, and mass spectrometry can offer both molecular weight and structural information on species in single particles very rapidly. TOF-MS can also be rather rugged and can be taken into the field.

My research focuses on aerosol particles between 0.1 μm and 10 μm. Bacteria, though most often seen with optical devices, fall within this size range where detection by mass spectrometry is ideal. Single particles down to 12 nm have been detected by mass spectrometry so that viruses, though considerably smaller than bacteria, can also be analyzed (see Figure D.5).

The instrument in my laboratory uses laser desorption ionization with a Nd:YAG laser and a TOF-MS. The particles are drawn into the instrument on a continuous basis and undergo a supersonic expansion when they pass through the inlet nozzle. During the expansion, the particles pick up different speeds that are a function of their size. They then pass through two scattering lasers. The time it takes the particle to travel between the two lasers can be correlated with particle size, allowing the particle size to be determined precisely. Knowing the particle speed and position, it is possible to time its arrival at the center of the spectrometer with a Nd:YAG laser pulse (266 nm). The pulse is able to desorb ionized species from the particle, which can then be analyzed by the spectrometer.

When ionized species are desorbed from the particle, care must be taken so that not every species is fragmented. Given a mixture of bacteria, fragmentation

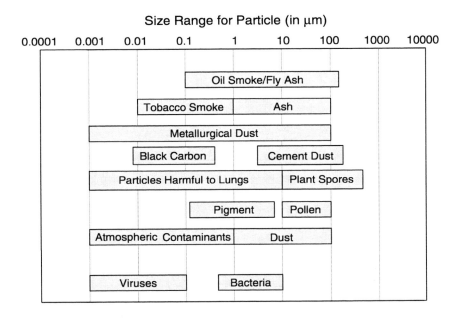

FIGURE D.5 The size of viruses and bacteria allow detection by mass spectrometry.

will cause all of the bacteria to look the same. However, gentle desorption and ionization using long laser (i.e., infrared) wavelengths, detection, and further fragmentation and detection in a tandem MS system can yield more detailed information in a useful manner. There are also technologies such as scanning mobility particle sizers currently available to allow particle size selection without using optical or aerodynamic methods. These methods enable size selection down to 3 nm.

Another important technology in the national security and homeland defense arena is ion trap secondary ion mass spectrometry. Many chemical warfare agents are not volatile and tend to condense on particle surfaces. Research at Idaho National Engineering and Environmental Laboratories has used this technology to analyze mustard agent on the surface of soil particles down to a surface coverage of 0.07 monolayers.

The development of matrix-assisted laser desorption ionization (MALDI) has advanced the entire field of mass spectrometry. To use this ionization method, the sample is mixed into a matrix that absorbs the laser wavelength extremely well (approximately 10,000:1 matrix:analyte) and the mixture is placed on a solid substrate. Absorption of the laser causes the matrix to explode, ejecting the intact, nonvolatile molecules of interest into the gas phase. Proton exchange or alkali metal attachment occurs in the gas plume and the ionized species can be detected.

The technique does require sample preparation, and reproducibility is often poor, although the results can be coupled with other analytical techniques for more reliable sample identification.

Ideally, scientists would like to be able to perform laser desorption and analysis directly, but typical laser wavelengths cause fragmentation of bacteria and other particles. Due to the low energy produced by infrared lasers, however, bacterial fingerprints can indeed be obtained as shown by researchers at Lawrence Livermore National Laboratory. It is also possible to detect much larger species, an impossible task with earlier technology. Infrared laser desorption techniques are undergoing constant improvement.

It is desirable to have hand-held, lightweight, and rugged instrumentation to take into the field. Whether the operator is a scientist or a soldier, analytical technology must become portable. For time-of-flight mass spectrometry, this is an especially difficult challenge, as it is the long flight tubes that enable the high resolution to be obtained. However, Robert Cotter of the Johns Hopkins School of Medicine has developed an off-line TOF-MS instrument with a flight tube that is only 3 inches long. The technique, coupled with MALDI for off-line analysis, is successful when used for particles of high mass, such as spores. Spores and other heavier particles travel more slowly, so the shorter flight tube is sufficient to obtain measurements for analysis.

In conclusion, there have been advances in the recent past in both on-line aerosol technology and off-line analytical techniques that show great promise for activities such as the analysis of spores. The next logical step is for chemists and chemical engineers to merge the two types of technologies.

NEW APPROACHES TO DECONTAMINATION AT DOE

Mark D. Tucker
Sandia National Laboratories

Today I will offer you information on the decontamination efforts in the Department of Energy Chemical and Biological National Security Program. In doing so, I will give a brief background of decontamination, touch on some key elements of decontamination, and present some of my own ideas on the future directions and critical needs for decontamination.

There are two main activities related to the decontamination of a site infected by a chemical or biological agent. A first response by firefighters, HAZMAT units, the National Guard, and other emergency personnel cannot be fully planned for and has the objective of rapidly achieving decontamination levels safe enough for treatment and evacuation of casualties. First responders are not attempting to obtain an agent-free site in every corner and crevice. Facility restoration is the second main decontamination activity and is accomplished by a variety of government agencies and possibly private industry. Facility restoration is a planned and controlled event; the amount of damage incurred by the decontaminating agent is minimized.

In the past, decontamination efforts were focused on neutralizing or killing chemical and biological agents, while little thought was given to reusing equipment at an attack site or to environmental effects. This led to the development of very highly toxic and corrosive formulations to neutralize agents such as supertopical bleach, a very toxic and corrosive mixture of sodium hydroxide and calcium hypochlorite. In the last five or six years, a new decontamination concept has emerged that tries to save and reuse equipment and restore facilities. This requires the decontamination formulations to be more environmentally and people friendly, in other words nontoxic and noncorrosive. Many of the historic decontamination formulations that are still in use today are being replaced.

The Department of Energy/National Nuclear Security Administration's Chemical and Biological National Security Program was initiated in fiscal year 1997 with the objective of developing technologies and methodologies to respond to domestic terrorist attacks that use chemical and biological agents. Of the four main thrust areas in this program, the first is detection and has the goal to develop technologies to detect both chemical and biological agents. The second thrust, biofoundations, is a basic science program that analyzes characteristics of microorganisms in order to develop better biological detectors. The third thrust uses modeling and simulation to examine the fate and transport of chemical and biological agents. Modeling and simulation can offer information on how far an agent will disperse and where restoration efforts should be focused, given, for example, an attack in a subway station. The final thrust is decontamination and remediation, which aims to develop effective and safe (nontoxic and noncorrosive) formulations and technologies to rapidly restore civilian facilities after a terrorist attack. One simple but elusive goal is to create a single formulation to be used by first responders that destroys all chemical AND biological agents. A second objective is to address the complex issues associated with final decontamination and remediation of a facility including "how clean is clean" and convincing the public that a facility is safe.

Three types of projects are funded under decontamination and remediation: methodology (deployment and use), new nontoxic formulations and technology, and field verification. In the area of methodology, two projects, one at Oak Ridge National Laboratory and one at Lawrence Livermore National Laboratory, are focusing on how clean is safe. Several nontoxic formulations have been developed, including the Sandia Decon Foam, which is active against chemical and biological agents, and the L-Gel, which travels through air ducts. A plasma jet technology for decontamination of sensitive equipment has also been built. Mock offices at the Dugway Proving Ground were contaminated with anthrax surrogate, and four novel emerging decontamination technologies were tested there as part of the field verification and testing program. Data from this test was used for technology evaluations and selections for the decontamination and remediation of the Capitol Hill office buildings.

I have been working on the Sandia Decon Foam formulation (see Figure D.6). Generally, foam is quicker to deploy and react with chemical and biological agents than a water- or fog-based decontaminant, and it has very low logistic support and water demand. The Sandia Decon Foam has very low toxic and corrosive properties and provides a single decontamination solution for both chemical and biological agents.

Decontamination of the Capitol Hill office buildings was a learning experience for our nation. The Sandia Decon Foam was one of many decontaminants used in the cleanup efforts. It became clear that no single technology is universally effective and that a suite of decontamination technologies is needed. It is also necessary to develop better methods to decontaminate sensitive equipment and objects such as electronics, computers, copiers, and valuable paintings. Consequently, facilities and standard methods are required for product testing.

Decontamination programs exist in other federal agencies as well. For example, the U.S. Army Soldier and Biological Chemical Command in Edgewood, Maryland, is focusing on military needs. They are looking at solution, oxidation, and enzymatic chemistry for liquid-type decontamination formulations, especially decontamination of sensitive equipment. Their developing technologies include supercritical carbon dioxide, solvent wash systems, thermal approaches, and the plasma jet, which is a cooperative project with the Department of Energy. They are also conducting large-scale tests with solid-phase chemistry focused on decontamination of bulk quantities of chemical and biological agents, if an entire military base, for example, needed to be decontaminated. Most of these types of projects are funded by the Department of Defense.

FIGURE D.6 Sandia Decon Foam is deployed.

In conclusion, I think future work should focus on several areas of decontamination. First, we need noncorrosive decontamination formulations and technologies for use on sensitive equipment and items as well as methods to decontaminate hard-to-reach places like air ducts. Fundamental issues like the number of anthrax spores necessary for infection and how clean is clean should be investigated. We also need more extensive lab and field data, and large-scale testing of decontamination formulations and methodologies, especially coordination and integration with sensors and modeling and simulation. Ideally, if we could tie all these things together, effort could be directed at the problem areas without wasting time on areas that remain clean. Personnel decontamination is also a main issue. With each anthrax incident, real or perceived, a personnel decontamination effort goes into effect, putting people through portable showers and spraying them with bleach, for example. This may not be a realistic measure as shown by the Tokyo subway incident: almost everyone who went to a hospital hopped in a cab and went on their own, thus contaminating the hospitals. Sociological issues are important, as well. First responders destroy evidence when they deploy foam; coordination with forensic and criminal investigators must occur. What evidence will be admissible in court? Finally, regulatory issues will also come into play in establishing standard test methods to determine the efficacy of and give approval to sensors and decontamination formulations for chemical and biological agents.

HOW INTEGRATION WILL MAKE MICROFLUIDICS USEFUL

Stephen R. Quake
California Institute of Technology

You've already heard about the basics of microfluidics and the lab-on-a-chip concept from Andrea Chow. So today, I'm going to talk about the gadgets based on microfluidics that my research group is making. I hope that this talk sparks your imagination and helps you to envision ways that microfluidics can help in the area of national security and homeland defense.

Over the last 10 years or so, scientists and engineers have been focusing on miniaturizing specific functions of lab equipment rather than miniaturizing the entire lab. This is because they had no way to integrate all of the pieces; the plumbing was lacking. More specifically, plumbing is controlled by valves, which are extremely difficult to miniaturize. To solve this problem, several years ago my research group developed a soft microvalve system using multilayer soft lithography.

If you have two layers of orthogonally arranged pipes and fill the bottom layer with fluid, a pneumatic pressure applied to the top pipe will cause the bottom pipe to pinch closed (see Figure D.7). This creates a valve. The valves are made of a soft, inexpensive polymer that allows the use of low actuation forces on the valves and yields small footprints. Pumping in this system is based on peristalsis, enabling the movement of multiple nanoliters per second through valves 100 μm wide and 10 μm deep. A nanopipette is built into the system, with different size plungers allowing different amounts of fluid to be introduced.

A number of devices have been made based on this microvalve system. One of the first devices built was a cell sorter on a chip. By manipulating nanoliters of fluid, different strains of fluorescently activated *E. coli* were introduced, sorted according to their fluorescent properties, recovered from the chip, and cultured.

My research group felt that high-throughput protein crystallization on a chip would be a useful challenge to meet. We first determined that it was possible to grow protein crystals on a chip, but only under the same conditions that allowed crystal growth in bulk experiments. The next experiment screened for protein crystallization, a difficult process. Success seems to be more due to sheer numbers and probability than on rational design, and due to the small amount of protein available for use, mixing and metering during the screening process are difficult to achieve. The microfluidic, microvalve system we have developed is able to overcome these obstacles.

The mixing ability of microfluidic systems was tested using a rotary pump, a circular channel with inputs and outputs that can be peristaltically pumped, opened, and closed. It was found that after only a few minutes of active mixing (due to pumping), a uniform mixture of particles is obtained that would have taken hours to achieve by diffusion. This is also useful for accelerating diffusion-

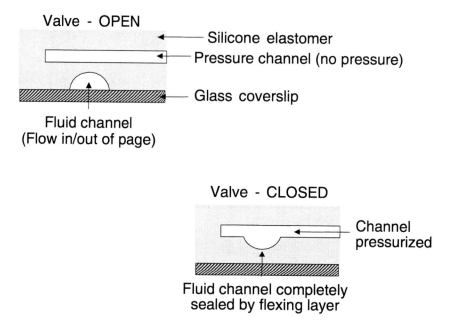

FIGURE D.7 Integrated microfluidic devices, which contain valves that can be pinched closed by pressurized channels, are fabricated from rubber by soft microlithography at low cost.

limited reactions; a 60-fold enhancement in the kinetics of some assays has been measured. The rotary pump has also been used as a 12 nL polymerase chain reaction system.

In the electronics industry, a Pentium computer chip has hundreds of millions of transistors and only a hundred pins in and out. If each transistor had to be addressed individually, it would be impossible to have such a chip. Microfluidic systems are similarly easy to control: n fluid lines can be controlled by 2log n control lines. Additionally, the pressure to actuate a valve depends on the width of the control line, so by choosing our pressure carefully, one thinner fluid line can be closed while the wider fluid lines remain open due to insufficient pressure. This idea has been used to develop microfluidic systems that can screen enzymatic libraries and perform in vitro transcription translation of DNA to protein in approximately 30 minutes.

I am sure that this technology is useful in many ways to chemists and chemical engineers and can help in the arena of national security and homeland defense.

BIOSYNTHETIC ENGINEERING OF POLYKETIDE NATURAL PRODUCTS

C. Richard Hutchinson
Kosan Biosciences

This morning I'd like to address basic medicinal chemistry, and specifically, how to manipulate drugs from natural sources. Biosynthetic engineering has recently been exploited in antibacterial drug discovery, but can also be applied to anticancer drugs, antiviral drugs, and others.

Kosan Biosciences was formed almost 6 years ago, founded on an interest in polyketides, microbial metabolite-based drugs. Polyketides have many diverse chemical structures including erythromycin, which will be mentioned again later. These chemicals include fused-ring aromatic compounds, compounds decorated with sugars, and compounds with large stretches of double bonds. Each of these compounds has different biological activities and utilities, but they are all made in nature by very similar biochemistry.

This biochemistry resembles how long-chain fatty acids are made. Acyl-coenzyme A substrates can be carboxylated, reduced, dehydrated, or otherwise changed from the original molecule. Each of these biochemical activities is accomplished one at a time, by single-function enzymes that are produced by an organism and collectively organized to make a particular kind of long-chain fatty acid. Polyketides use enzymes like this in much more complicated ways than required to make fatty acids. While biological organisms may use only a handful of activities, with modern genetic methods biochemists can combine genes for any or all of the activities to create a complex array of thousands of natural products.

Imagine a set of genes in the chromosome of a bacterium. The protein products of these genes have individual active sites (domains) and are grouped in modules. Each module can have a different number of active sites and hence a different function: joining bonds, selecting and loading substrates, and carrying the substrates, for example. Not every module is used for every synthesis; this characteristic imparts the synthetic flexibility and diversity that creates the eventual polyketides. Modules are strung together in very large proteins to take simple substrates the organism provides, assemble them in a sequential manner, and produce a polyketide. The enzyme then continues to produce polyketides and accumulate the product (see Figure D.8).

For Kosan and others in the pharmaceutical industry, the intent is to learn enough about these enzymes from a structure-function viewpoint so they can be manipulated. Because polyketides are built by starting from one point and continuing sequentially along the pathway dictated by protein structure, a biochemist can trace through a molecule and make structure-function predictions for the assembly enzymes with reasonable accuracy. Using recombinant DNA methods,

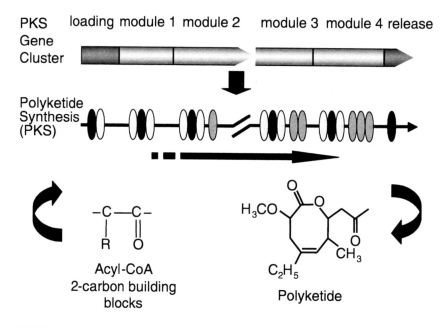

FIGURE D.8 To synthesize polyketides, an enzyme joins protein modules and is recycled to continue molecule production.

it is possible to remove a single module or a specific domain within a module, place it somewhere else in the protein, and make a product with a different structure and functionality.

This technique has been used in the attempt to create new antibacterial agents that target drug-resistant pathogens. For example, erythromycin is a very well known antibacterial drug that has been used for approximately 50 years for a variety of infections, primarily in the lung. A polyketide synthase produces erythromycin and has six modules, each with a certain number of active sites. By manipulating the active sites or modules of the polyketide synthase, the alkyl groups could be taken out or changed, the oxidation state of the hydroxyls or the ketone could be changed, or the length and size of the lactone ring could be changed. Although many variations of the synthesis were attempted, none developed a better antibacterial drug than the existing erythromycin. The experiment was valuable, though, because it established a paradigm for structure-function relationships.

Similarly, we could create a mutant in part of the polyfunctional protein and allow the synthetic substrate to be accepted by the functional parts and carried through to produce a completely different compound. The benefit of this type of manipulation is that synthetic procedures that are very difficult for chemists to do can be accomplished easily. These processes have allowed scientists to test a large number of macrolide antibiotics and to develop lead compounds for a new type of drug called ketolides, which has activity against usually drug-resistant pathogens.

To summarize, polyketides are simply bacterial products that include erythromycins and chemically derived ketolides. New erythromycins and polyketides can be made by genetic engineering of the polyfunctional giant proteins called polyketide synthases or by taking products of the engineered microbial metabolism and further modifying them. These new antibiotics can be more insightfully designed due to increased knowledge of structure-function relationships for how such antibiotics bind to bacterial ribosomes. This allows treatment of drug resistant organisms that are of importance in community-acquired infections, pathogenic diseases, and especially unique bacterial pathogens that have previously been ignored because they had never been a threat. Although we cannot predict drug efficacy, drug production is more approachable using the technology of microbial metabolism modification and structure-function information for bacterial ribosomes.

CHALLENGES IN RAPID SCALE-UP OF
SYNTHETIC PHARMACEUTICAL PROCESSES

Mauricio Futran
Bristol-Myers Squibb

Even though synthetic organic chemistry is a very old scientific pursuit, I believe that today most medications needed for homeland defense will be created by means of synthetic organic chemistry. This type of research and development is done by industrial organic chemists and chemical engineers, and there are many opportunities to involve academic scientists to increase the speed of development.

The pharmaceutical industry can improve by better partnering with academia to focus on solving the most pressing problems. In general, there is a lack of understanding of automation and parallel experimentation among chemistry graduates. Engineering students also need training in modeling and computation, specifically in the areas of predicting physical and transport properties, computational fluid dynamics, specific process operations, and control theory. This facilitates obtaining scale-up parameters from small-scale experiments better known as "micro-piloting," which is of utmost importance. Further development in the areas of in-line analytical and microchip technology would be useful for rapid drug development.

In a national emergency, the pharmaceutical industry could be called upon to bring a medication to market, to a commercial scale, very quickly. Normally, the complex scale-up process does not operate on such a short timeline. Assuming that we can partner with government agencies to ensure compliance in a streamlined manner, assuming we have better predictive toxicology, and assuming that the development of the dosage form (capsule, tablet) is not an issue, there remain a number of barriers to the scale-up process to create the active pharmaceutical ingredients.

Often in a medicinal laboratory, synthetic routes are linear instead of convergent and may contain many steps with low yields and operating conditions not favorable to scale-up (for example, strong exotherms, mass transfer limitations, and gums/heavy slurries). Reactions involving chiral molecules may have low selectivity for the medicinally active isomer. In addition, the reagents used may be toxic, and using large amounts of toxic agents for a scale-up is not an acceptable solution. Some laboratory materials are not readily available in large quantities.

Scale-up to 100-g quantities for animal testing, which must precede human studies, typically takes 3 to 8 months. Then, production must be scaled up to commercial quantities. Traditionally, the entire process takes 2 to 3 years—too long to be useful in a national emergency.

Much of the knowledge needed for manufacturing a pharmaceutical is related to the last step in the process, including understanding the bulk active ingredient, its impurity profile, chiral purity, and crystal form. To cut down the entire scale-up process to only 6 months would require some of this knowledge to be investigated from the beginning, in parallel with other information. A 6-month timespan may still seem slow, but that is one-fourth the time normally taken by industry.

The scale-up process can be broken into smaller steps. First, the synthetic pathway must be defined. This includes determining how to make the final compound in a safe, environmentally sensitive, and affordable manner. This may require thousands of experiments and an equally large amount of analytical tests. The second step involves developing the pathway into a process. Issues in this stage include ensuring the right chiral and chemical purities, which requires hundreds of experiments. The last stage is performing hundreds more experiments to learn about process parameters and optimization (see Figure D.9).

To minimize the time required for these three steps, chemists need to appreciate the process-related issues not only of how this elegant structure can be made, but also how it can be made efficiently. They should also know chemical engineering concepts like kinetics and continuous processing so that these ideas can be incorporated into the synthesis process. Chemists must be comfortable with automation—with the tools they need for the job. Commonly, automated systems are technologies that sit underutilized in chemical laboratories. Chemical engineers are lacking in-depth knowledge of organic chemistry and particular spectroscopic methods, which are essential in modern industry. It is impossible to function in the pharmaceutical industry without understanding what the chemists are saying. It would also be advantageous for engineers to have hands-

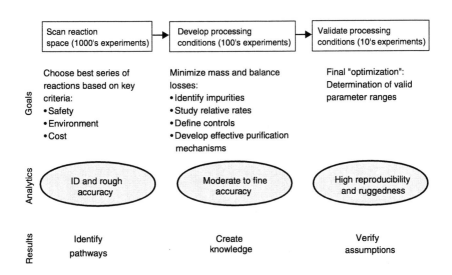

FIGURE D.9 Each stage of drug development involves many issues and experiments.

on industrial experience with taking a product from the concept stage to the commercialization stage. The educational system can address many of these problems.

As mentioned earlier, parallel and automated methods are a key part of making the scale-up process faster. Automation technology already exists to provide good agitation, heating, reflux, inert headspace, sample filtration, and sample taking in a variety of vessel sizes. However, there remain challenges in the automated handling of solids and in the use of on-line spectroscopic instrumentation on very small samples, which would aid the quick determination of process optimization, parameters, kinetics, partition coefficients, and the like. It is essential that the automated technology be user friendly, so that chemists can focus on the chemistry and not on how to use the equipment.

Continuous systems will also become an important topic for rapid scale-up during a national emergency. First, continuous reactions allow the scale-up to be considerably smaller than that needed for a batch reaction and allow equal or greater productivity. Continuous systems can much more easily handle highly reactive, unstable ingredients. They can also better handle hazardous reagents; since the reagents are made in situ and are used immediately, they never accumulate.

Use of continuous systems will require chemists and chemical engineers to work together more closely. It will also need more and improved prior knowledge of the physical properties and transport properties of reagents. Consequently, our predictive and modeling capabilities must improve.

Currently, after each step in the synthesis process, a sample is analyzed to approve or reject the operation. Not only is this inefficient for the industry in general, this is also a stumbling block for a rapid scale-up. Controls and measurements of the process need to be accomplished in situ, which is again an issue that requires the chemists and chemical engineers to work together toward a solution. It may, in fact, involve lab-on-a-chip technologies and microreactors that were presented by other speakers in this workshop.

To integrate everything into a continuous process, accelerated reaction kinetics are needed. A reaction that takes 20 hours to complete is difficult to use in a continuous process. However, we have microwave chemistry, ultraviolet chemistry, laser stimulation, sonication, and other technologies to accelerate proven but slow reaction chemistries. Research in this area is merely beginning.

Another area that is lacking is the chemistry of solids. In pharmaceuticals, an active drug is often made into a tablet or some other solid. Unknowns remain in the areas of solids flow, solids compressibility, and compaction. The physical incompatibility of various inert ingredients also remains a mystery. The engineering community has a great opportunity to help in this area.

In conclusion, research is needed for the improvement of automated tools, handling of solids, modeling and computational tools, and in-line analytical technology. To speed the scale-up of pharmaceuticals in a national emergency, these issues must be addressed by both chemists and chemical engineers.

E

Results from Breakout Sessions

Workshop participants were broken into the Red, Green, and Blue groups, with efforts made to ensure that participants from academia, industry, and government were evenly distributed among the groups. After the formal presentations, the participants were able to meet in these small groups to discuss chemistry and chemical engineering discoveries, challenges, technical barriers, and research needs. The four topics set forth in the workshop task statement—discovery, interfaces, challenges, and infrastructure—were covered in each of the breakout sessions as they intertwined with the session topic.

Participants were asked to offer short phrases describing chemistry and chemical engineering's contributions or needs in each of the breakout sessions. Then, the most important subject areas were chosen by voting. The following list of subjects has been taken from each breakout group's reports of their discussions to all the workshop participants; the number of votes an item received follows that item.

Discovery—Red Group
1. Polymerase chain reaction (PCR) and DNA technologies (18)
2. Microanalytics (15)
> optical storage media, scanning probe microscopy, IR sensing, miniaturization, microfluidics (1), forensics (2), molecular recognition (2), fluorescence resonance energy transfer assays (4), biological matrix assisted laser desorption ionization mass spectrometry (3), chemical signature detection (3), single particle monitoring, atmospheric reaction of organics, nanoanalysis, whole cell assays, sensor arrays (1), fiber optics, low photon optics, metal detection, NMR imaging, optical methods

99

3. Electronic materials and processing (14)
4. Stealth technology (13)
5. Remote sensing and analysis (12)
6. Antibiotics (10)
7. New polymers and composite materials (10)
8. Other

> chemically amplified photoresists, biological adhesives, nuclear safe-guard materials (1), reactive armor, energetic materials (1), personal protection materials (3), packaging materials (food), high-temperature materials, kinetic energy penetrators (1), nanoparticles, lighter and stronger materials (1), ceramics, chemical additives, synthetic membranes, high-performance fibers, fluoro polymers, coatings, proliferation-resistant processing (1), advanced batteries (6), plutonium management (3), fuel cells (1), photovoltaics, catalysis (2), supercritical processing, combinatorial methods, chiral synthetic methods (1), nuclear energy, new separation and purification techniques (2), chemical transport models, computational chemistry (4), informatics (2), modeling and simulation (4), education (4), chlorofluorocarbon replacements, new synthetic methods, chemical vapor deposition microelectronic processing (1), electrophysical monitoring, crop protection chemicals (1), pharmaceuticals, DNA vaccines, methane as feedstock (1), water treatment (1)

Discovery—Green Group

1. Materials (37)

> colloidal nanoparticle technology (e.g. aerosol generation) (4), nonlinear optics switches, biodegradable surfactants, biocidal surfaces, controlled release/delivery and encapsulation for delivery systems (4), development of personal protection systems, fuel cells/batteries (6), imprintable polymers/zeolites (1), supercritical fluid technologies, development and application of photoresists and semiconductor synthesis (12), high-performance composites (structural components) and nanocomposite materials (11)

2. Analytical detection (33)

> application of advanced spectroscopic techniques for characterization of volatile/semi-volatile particulates in the atmosphere (5), diode lasers (2), single organism detection (1), emissions testing, single molecule detection (3), polymerase chain reaction (6), electrochemical sensors (1), high throughput screening (3), remote sensing (4), real-time analytical techniques, diagnostics of biomeds, aerosol detection (2), drug testing, advanced detection/electron capture detection, automation of chemical analysis, mass spectrometry of macromolecules (6), multidimensional fluorescence spectroscopy

3. Synthesis (22)

   genetic engineering (5), combinatorial synthesis (1), production of raw materials from biomass (1), developments in catalysis (10), new biocides and delivery systems (5)

4. Computational methods (20)

   protein-protein inhibitor design (2), computational chemistry, risk assessment (1), wireless communications (1), complex environmental chemistry (1), computation of fluid dynamics (1), better modeling techniques (7), information systems for wide area screening, genomics/biological information (7)

5. Separations (4)

   new chemical/nuclear remediation technology (1), phage display of antibodies, breathalyzers(1), water/air filtrations technology, microfluidics, microseparations for analytical applications (1), capillary electrophoresis (1)

6. Other

   high performance materials synthesis/processing, semiconductor synthesis/processing, drug discovery/development, detection systems, developments in catalysis, stockpile stewardship, supply chain modeling/just-in-time synthesis, rapid translation, better risk management techniques (1)

Discovery—Blue Group

1. Analytical methods—spectroscopic methods (25)

   laser spectroscopy, scanning probe microscopies, surface chemistry, multidimensional multipulse nuclear magnetic resonance, remote sensing, chemometrics, statistical analysis

2. High throughput biological analysis (18)

   polymerase chain reaction, DNA sequencing, genomics, proteomics, cell sensors, neuron sensors, flow cytometry, single cell/single molecule detection, ion mobility, high resolution mass spectrometry, nonvolatile mass spectrometry, miniature mass spectrometry, nuclear magnetic resonance imaging, fluorescence detection and imaging

3. Materials science (17)

   high energy materials, liquid crystals, composite materials, nanoscale science, smart materials, selectively permeable materials, micromachining, new materials, self assembly

4. Materials synthesis (14)

   thin-film coatings on optics, gun powder and explosives tagging, solid-state lasers, light-emitting diodes, magnetic data storage, optical data storage, radar-damping polymers, fiber optics

5. Mechanism theory modeling (12)

   state-selective chemistry, mechanisms and tracing of isotopes, atmospheric chemistry, actinide chemistry, free-radical processes and remediation, understanding biosynthetic pathways, computational chemistry (quantum mechanical molecular modeling), computerized algebra, high throughput screening, combinatorial analysis

6. Molecules (organic) (11)

combinatorial molecular biology and DNA shuffling, new antibiotics and antivirals, selective complexation and recognition chemistry, new polymers, biomaterials, combinatorial synthesis

7. Other

separation chemistries and preconcentration methods, high-performance materials, synthesis/processing, semiconductor synthesis/processing, drug discovery/development, detection systems, developments in cataly-sis (biocatalysis, enantioselective synthesis/catalysis), solid state (micro sensors, microbalances, macromolecule probes, surface acoustic wave devices, array detectors, photonic/mechanics—optical tweezers), energy (fuel cells, battery technology), methods (internet sensing, remediation and reprocessing of nuclear materials, robotics, microelectronics and chip processing, microelectromechanical systems, bioremediation, quality control, rapid identification of biotoxins, microencapsulation, lab-to-market/rapid scale-up)

Grand Challenges—Red Group

1. Sensing and detection (23)

materials; remote sensing; nanofabrication; protein-surface interactions; low-cost, real-time methods; species-specific databases; portable, species-specific detection of chemical and biological weapons; mass spectrometry libraries for chemical and biological weapons; (un)masking chemicals/explosives; ports and harbor security; methods for efficient micro- to nanofabrication

2. Energy (22)

hydrogen fuel cell chemistry, hydrogen generation and storage, hydro-gen distribution, alternative energy, energy/feedstock optimization, methanol chemistry, photovoltaics, propellant and combustion chemis-try, nuclear materials—plutonium and nuclear waste

3. Countermeasures (15)

presymptomatic diagnosis, real-time detection, "hate-detector"— chemical signals of personal behavior, personal/surface decontamina-tion, nonlethal weapons, tranquilizing chemicals, personal protection materials, materials for selective decontamination and absorption, medical countermeasures, drug targets, smarter bombs

4. Computational methods (10)

modeling, simulation and prediction, materials (harder, stronger, lighter), combustion, catalysis, bioinformatics, molecular interactions, chemical and biological weapon dispersion, stockpile stewardship

5. Collective protection (8)
   buildings/transportation systems (filtration, indoor air quality, structures, materials)
6. Political (7)
   education—real and implied risks, education on decision making, nonproliferation outreach, chemical and biological weapon treaties, industrial and political reconciliation, chemical and biological weapons intelligence
7. Domestic risk abatement (industry) (5)
   safer and alternative chemicals and processes, inventory control of precursors

Grand Challenges—Green Group
1. Advanced sensing technologies (20)
   biological sensors (2), chemical sensors (6), improved cost and reliability of sampling (12), advanced lasers, miniaturization
2. Understanding biochemistry of agents (19)
   sporulation/germination, generic toxicity detection, pre-symptom diagnosis (2), antivirals and new antibiotics, safe immunization techniques, genomic and proteomic data on agents (1), cures for addiction (2), food safety
3. Energy independence (17)
   cheap, clean domestic energy (9); fuel cells for vehicles (1); reliable portable energy (4); lean NO_x catalysis (1); C_1 catalysts; nuclear waste disposal (1)
4. Detection strategies (17)
   ruggedized field equipment (6), instantaneous detection of threat agents (4), integration of sensing and response (2), automation of analysis (2), unambiguous identification of threat agents (1), remote sensing (2), detection and interdiction of illegal drugs
5. Mitigation of threat agents (14)
   nerve agent response, how clean is clean, biocidic clothing and coatings, blood processing, storage, and substitutes (1), detection/interdiction of illegal drugs
6. Systems integration (14)
   predictive modeling, transport and fate of threat agents, large-scale medical surveillance, performance verification and standards
7. Political
   unambiguous identification of people (1), new weapons systems, treaty verification (2), nuclear and radiochemical education (1), education of scientists and the public, education of public regarding hygiene (3), understanding risk perception and communication

Grand Challenges—Blue Group

1. Universal threat detection

   peptide equivalent polymerase chain reaction, real-time molecular inventory, electronic dog's nose, function-based detection, signal-to-noise challenges, sensor array city (universal monitoring), chemical reactivity at molecular level

2. Universal threat neutralization

   "Guardian Angels", rapid detoxification, molecular machine countermeasures (virus model), less explosive fuels, universal antidote (for antiviral, antibacterial, and chemical toxins), standards for testing to known capabilities, construction issues, preclinical diagnosis

3. Global energy security

   fuel cells, catalysis/precious metals, energy storage capabilities (batteries)

4. Universal threat reduction

   energy absorbing structural materials, statistically designed instrumentation, environmental sustainability

Barriers—Red Group

1. Crosscutting issues

   access to internal intellectual property (9), national center for biological and chemical weapons testing (5), lack of materials data (4), lack of profit incentive (3), translation of concept to product (3), collaboration of industrial concerns (1)

2. Energy

   nuclear waste management (6), methane to chemicals (2), nuclear waste reprocessing (2), energy storage (2), small research base in catalysis (1), long-term use of biomass (1), cleaning up fossil emissions, high-temperature superconductors, electrode redox kinetics

3. Sensing and detection

   standard test methods for detection (5), high throughput screening (1), label free detection (4), geographical and temperature variability (2), better microfluidics

4. Countermeasures

   new decontamination chemistries (6), transport issues

5. Computational methods

   user-friendly computational tools (4), code optimization (3)

6. Collective protection

   filtration for biological and chemical agents (2), underwriting performance testing

7. Political

   supply of chemistry and chemical engineering graduate students (6), interdisciplinary training (2), curriculum does not address current problems/needs (1)

8. Domestic risk abatement
   life-cycle assessment tools (1)

Barriers—Green Group

1. Sampling (19)
   sampling artifacts, environmental interferences and definition of background, sample collection and concentration, impact of environmental dispersal, sampling methods pre- and post-decontamination
2. Biochemical
   early host-pathogen interactions, relationship of genes to protein function, clinical trials of therapeutics, integrated genome database, identification of problems with therapeutic agents, toxicity threshold of chemical and biological weapons, antibiotic resistant strains and modified organisms, combinatorial nanoscale analysis, relation between bioactivity and concentration
3. Detection (14)
   improved heat exchangers, light batteries, limit of detection for molecules and cells, room temperature detectors, more efficient coolers, low-cost and real-time particle identification, chemical contact detection, landmine explosive detection, developing methods for creating chemical and biological agent libraries, origin of threat—chemical signature, protein surface interactions, container inspection (closed, rapid), chemical and biological sample preparation, chemical and biological weapon test standards, integration
4. Structure-property relationships (12)
   rational design of materials and molecules
5. Logistics (12)
   location of sensors, response equipment, response personnel, communications, and vaccines; information management (dissemination and control); definition of new threats
6. Computational speed and storage (5)
7. Efficient separations (7)
8. Realistic field testing (7)
9. Energy (9)
   new catalysts for diesel aftertreatment, technology for nuclear waste, long-term carbon sequestration

Barriers—Blue Group

1. Energy-related limitations (13)
   efficient sunlight capture (1), understanding climate change, efficient energy storage (5), limited energy resources (2), need good nuclear reactor and waste management, dependent on limited resources (5)

2. Cultural and communication barriers (11)

   knowledge of limitations of current detection technology (1), classified information, cross-discipline compartmentalization (1), lack of peer review (1), utilization of talent (7), lack of students (1), needs are not adequately defined

3. Sensors (11)

   sensors to systems (2), transduction from chemical to optical/electronic signals (6), primitive nanoscale fabrication methods (3)

4. Biological issues (11)

   reagents for biological assays (2), testing protocols for antivirals, insufficient infrastructure for class 3 labs (1), understand molecular recognition (5), structure function understanding (2), understanding viruses, identification of drug targets, cell-specific (targeted) medicines, computational chemistry (1)

5. Funding (9)

6. Intellectual property issues (9)

   inexperience in technology transfer (5), intellectual property fragmentation (3), commercialization risk aversion (1), lottery mentality

7. Lack of long-term focus (3)

8. Communication

   public education (3), biotechnical and cultural barriers

Research Needs—Red Group

1. Basic science/understanding (24)

   protein folding, molecular basis of toxicity (2), catalysis (1), aerosol chemistry (4), molecular targets of agents (3), interfaces (1), molecule-surface interactions (12)

2. Sensing and detection (21)

   miniaturization (3), see-through sensing (4), remote sensing (10), new bioprobes (2), fundamentals of chemical and biological weapons detection (2), remote detection of emotional state

3. Energy (17)

   fuel cells and photovoltaics (8), closing the nuclear fuel cycle (7), alternate energy (2), new ionic conductors (1), batteries and energy storage

4. Materials (15)

   multifunctional materials (7), bioactive pharmaceutical agents (4), high-energy-density materials (2), high-performance textiles (2), biomimetic materials, processable materials

5. Methods to deal with emerging threats (11)

6. Measurement and prediction of physical/chemical properties and related database (4)

7. Improved decontamination methods (4)

8. Synthesis (4)

    new synthetic methods (4), genome synthesis, molecular evolution

9. New molecule process scale-up (3)

Research Needs—Green Group

1. Detection and identification

    light interaction with matter (10), spectroscopy, single molecule analysis (2), molecular recognition (2), microsensors (2), electrochemistry (e.g., microarrays) (1), more rapid identification of causative agents, techniques for personal detection, very small microelectronics

2. Interfacial chemistry

    efficient separations (11), basic understanding of interfacial catalysis, interactions of surfaces and threat agents, colloidal sciences

3. Biochemistry

    disease transmission (6), protein folding (3), basic immunochemistry (2), basic understanding of enzyme catalysis (1), generation of biomolecular products, biofilm badges, sporullation inhibition, source tracking of viruses/bacteria/chemical weapons, K-9 olfactory system, biochemistry of early host/pathogen interaction

4. Aerosol chemistry (7)

    atmospheric chemistry (2), environmental fate of threat agents (1), basic combustion chemistry (explosives)

5. Alternative energy sources (4)

    nuclear/actinide chemistry (4), identification of class 3 materials

6. Structure/function relationships (3)

    self assembly (3), basic research in microelectromechanical systems (1), database of structure relationships

7. Materials development (2)

    nanomaterials for molecular selection (2)

8. Other

    remediation of threat agents (1), nonlethal weapons (1), identifying chemical weapons production, modeling (2), design of bulk properties from electrical properties (1), process scale-up modeling (3), improved understanding of basic properties, new and better tools, molecular mechanisms of viruses

Research Needs—Blue Group

1. Advances in tools (12)

    methodology for establishing standards, structure function relationships, stabilized biologicals, chemical/biological amplification approaches, fundamentals of chemical engineering (scale-up issues, novel unit operations), high throughput sample analysis

2. Detection/sensor development (10)

>   biological processes for remote sensing, ultra-rapid DNA and RNA analysis, detector response analysis, exploitation of and materials for nonlinear optics technology, detection of agents of threat, enhanced spectroscopic and spectrometric resolution, data analysis and computational techniques for imaging

3. Material science (8)

>   scaling laws: structure/property relationships vs. size, fundamentals of collective behavior (biology, materials, polymers), nano- to macroscale-up, need focus on materials, need focus on solid state, need barrier materials, need chemically selective materials, energy-absorbing structural materials

4. Infrastructure (6)

>   national standards of detectors, government reagent repositories, centralized and "publicly" available research facilities, government coordinated approach

5. Education (2)

>   new approaches to teaching chemistry, updated curricula

6. Dual-use technologies (0)

7. Remediation and recovery (2)

>   fundamentals of ionizing radiation interaction with matter, quantum mechanics and molecular modeling for rate constants and mechanisms of destruction, better understanding of free radicals in aqueous solution

8. Energy (0)

>   cheap solar collection devices, batteries, new routes to hydrogen generation, fundamentals of nuclear waste

9. Catalysis (1)

>   focus on inorganic chemistry, catalysis and precious metals replacement